CHANNELS OF
DISCOURSE

CHANNELS OF

TELEVISION AND

The University of North Carolina Press Chapel Hill and London

DISCOURSE

CONTEMPORARY CRITICISM

EDITED BY ROBERT C. ALLEN

© 1987 The University of North Carolina Press

All rights reserved

Manufactured in the United States of America

Library of Congress Cataloging-in-Publication Data

Channels of discourse.

Bibliography: p.
Includes index.
1. Television criticism.   2. Criticism.   I. Allen,
Robert Clyde, 1950–
PN1992.8.C7C48   1987   791.45'01'5   86-24908
ISBN 0-8078-1732-5
ISBN 0-8078-4176-5 (pbk.)

FRONTISPIECE

The ideal TV room, as envisioned by Bloomingdale's in 1949.

(Courtesy of Bloomingdale's)

CONTENTS

CHANNELS OF DISCOURSE

INTRODUCTION
TALKING ABOUT TELEVISION
ROBERT C. ALLEN

What is there to say about television? At times it seems that the entire culture revolves around the images and sounds that emanate from the television screen, that all talk is somehow television talk. Yet, paradoxically, given the centrality and ubiquitousness of television in American and European culture, it somehow seems that not nearly enough has been said about television: the manner in which its sounds and images are organized; its nature as a powerful economic and social institution; the curious relationship between ourselves and the worlds, both fictional and nonfictional, that project themselves into our living rooms. Perhaps it is the case that a phenomenon so pervasive as television usually remains invisible to critical scrutiny.

This collection of essays contains more talk about television. A few words about the nature of that talk are needed before you participate in the discussions that follow. These essays introduce some of the major strands of current critical practice as developed in literary, cinema, and cultural studies; they discuss how television might be defined as an object of study within these critical frameworks and provide examples of the type of analysis that might be produced as a result. The approaches differ from each other in many respects, and each focuses on a different aspect of our relationship with television. Indeed, I asked each contributor to emphasize the particularities of the approach he or she describes. Despite their differences, however, all of the approaches outlined here grew out of, were strongly influenced by, or were developed in reaction to the insights into language and culture provided by structuralist linguistics and the "science of signs" (semiotics) that developed out of it.

Although each of the following chapters speaks a somewhat different critical language when it "talks" about television, those critical languages belong to the same linguistic family. I am using the term "contemporary criticism" as a shorthand designation for this diverse family of critical languages: semiotics, narrative theory, genre theory, reader-response criticism, ideological analysis, psychoanalytic criticism, feminist criticism, as well as other branches of the family not included here (the discourse analysis of Foucault and the deconstructive criticism of Paul De Man and Jacques Derrida, for example). Thus contemporary criticism marks out a very general set of assumptions about both criticism and the object of critical analysis that sets it apart from traditional literary criticism on the one hand and, because the object of study here is television, from traditional mass communication research on the other. It might be useful to suggest here the nature of those assumptions, keeping in mind that they are shared, in varying degrees, by the specific critical approaches discussed in each chapter.

All of the approaches regard television as one of a number of complex sign systems through which we experience and by which we know the world. Perhaps it would be more precise to say that television represents an ever-changing point of convergence for those sign systems. In the light of initial insights provided by semiotics pioneers Ferdinand de Saussure and Charles Sanders Peirce, scholars have attempted to describe the operation of those systems, their interrelationships, and their determinative effect upon our knowledge of the world around us. The one question that runs through each chapter might be summarized as "how are meanings and pleasures produced in our engagement with television?" The apparent naturalness with which we understand the sounds and images on television might seem to render this question unnecessary. After all, no one had to teach us how to "read" television programs. But the naturalness of our relationship with television is illusory. Television, like the cinema, painting, or photography, does not simply reflect the world in some direct, automatic way. Rather it constructs representations of the world on the basis of complex sets of conventions—conventions whose operations are hidden by their transparency. Furthermore, despite the seemingly self-evident manner by which we are able to make sense of television, that ability is, in fact, a result of our having learned those conventions of television reading—even though we are usually not conscious of their operation, nor can we remember having been taught them. In recent years scholars have given particular attention to those forms of cultural production that would seem to be most automatic and "natural": the nineteenth-century novel, the photograph, movies, and, increasingly, television.

One of the guiding principles of contemporary criticism is, as Jonathan Culler has put it, that "if we are to understand our social and cultural world, we must think not of independent objects but of symbolic structures, systems of relations which, by enabling objects and actions to have meaning, create a human universe."[1] This emphasis on structure and system has led to a shift in focus in contemporary criticism from the individual work (of art or literature) to the codes and conventions that operate within the work and link it with other works and other sign systems. Put another way, the work has been reconceived as a site of intersection for a complex tangle of signifying practices rather than as a self-sufficient, independent "thing." This is not to say that under analysis individual works simply dissolve into the externally originating codes and conventions that govern their ability to be read (as a novel, a film, or television program). Each work does present itself as a new reordering of cinematic, televisual, or literary material and is, to a certain extent, experienced as such by the reader or viewer. But however distinctive, each work is always a re-ordering of already-existing codes, conventions, and materials.

Contemporary criticism's foregrounding of the codes and conventions at work across individual works (or "texts," as they will be commonly referred to in the following essays) and the inevitable circuit of reference set up between texts would seem to be particularly appropriate in the case of television. Our experience of television is usually not of isolated works but of chunks of time filled with multiple texts carefully linked together so that they flow almost unnoticed one into another. A commercial is followed by a network promotion for a future program, which is followed by a "teaser" for the episode of a series about to begin, which is followed by a segment of that episode, interrupted by another commercial, followed by a "newsbreak" that anticipates the late-evening news program, and so on. Because the economics of commercial television is based upon maintaining the largest possible audience over the longest possible period of time, television programming practice actively discourages the viewer from thinking of the schedule in terms of a sequence of isolated and unrelated programs. In fact, enormous amounts of energy go into covering over the gaps between programs and stitching each segment into the larger programming fabric.

Contemporary criticism has also led to a reconsideration of the role of the author in the production of art. The traditional notion of the author or artist as the ultimate and single source of meaning within a work is difficult to maintain once one acknowledges the complex network of codes, conventions, precedence, and expectations in which every work inevitably participates and over which the author has little, if any, control. As Pierre Macherey has argued with regard to literary authorship, "The author certainly makes decisions, but, as we know, his decisions are determined; it would be astonishing if the hero were to vanish after the first few pages, unless by way of parody. . . . His narrative is discovered rather than invented, not because he begins at the end, but because certain directions are firmly closed to him. We might say that the author is the first reader of his own work; he first gives himself the surprises that he will hand on to us, he enjoys playing the game of free choice according to the rules."[2]

Where better to observe the circumscribed role of the author in contemporary cultural production than in commercial television? Because of the technological complexity of the medium and as a result of the application, to television production, of the principles of modern industrial organization (mass production, detailed division of labor, etc.), it is very difficult to locate the "author" of a television program—if we mean by that term the single individual who provides the unifying vision behind the program. Producers might come up with the basic idea and characters for a television series, but they rarely are involved in the writing of individual episodes. Television writers usually work in teams, and their jobs are finished with the production of a script that conforms to limitations already laid down by the producers. A given series might well employ a number of directors, who are unlikely to have had any part to play in the scriptwriting process and whose directorial styles must by indistinguishable from each other.

Just as contemporary criticism has questioned the notion of authorial genius as the single source of meaning in an artwork, it has also questioned the artwork's ability to show us the world "as it really is." This is not just because the artwork is to some degree a product of the artist's imagination, but, more crucially, because all our attempts to represent reality are conditioned by language, culture, and ideology. As Saussure pointed out nearly a century ago, words acquire meaning by virtue of their positions within a conceptual system of similarity and difference and not through any direct relationship with reality. To anticipate Ellen Seiter's discussion of this insight in chapter one, the word "cow" means something to us because it marks out those qualities we file under the heading of "cowness." It is only through that conceptual category that we are able to link up the word "cow" with a bovine creature we encounter in the "real" world. Even a photograph represents a cow to us in a particular way, through the operation of specific codes and conventions. Furthermore, what we are referred to by that photograph of a cow is, once again, the conceptual category of "cowness" rather than a "real" cow. In other words, contemporary criticism assumes that we experience the world through systems of representation that, at the very least, condition our knowledge of the world and, some would argue, construct that world.

Criticism of television news frequently revolves around notions of "bias" and "objectivity." Framing the discussion by these terms obscures the fact that there is no "unbiased" manner by which television (or any other system of representation) can show us the world. For the contributors to this volume, the question is "How does television represent the world?" not "Does television give us the 'truth' about the world?" Many of the following essays are concerned with the conventions employed by television that give the illusion of immediate access to reality and truth.

The dream of early semioticians was to develop a science of signs, whose goal would be the discovery of the laws that govern all instances of meaning production. This vision of semiotic research modeled after science has been tempered, in more recent criticism, by recognition of the enormous semiotic complexity of even the seemingly most simple communication act and of the fact that we do not experience the "laws" of semiotics except in their employment in specific instances and within specific contexts. The relationship between signifier (word, image, or sound) and signified (the concept for which it stands) is slippery rather than stable. Roland Barthes speaks of the "play" of signification. Furthermore, as reception theorists have argued, meaning does not reside "out there" in words on a page or dots on a television screen, but comes about as a result of a confrontation between viewer and image, reader and text. These confrontations, which occur so frequently and spontaneously that we seldom notice them as such, all occur within particular historical, cultural, and institutional contexts— contexts that inevitably condition the production of meaning. In the spirit of post-structuralist humility in the face of the daunting task of grasping the ways we make sense of the world around us, the essays in this book do not attempt to explain television once and for all. Rather they open up some lines of inquiry into television and, in doing so, suggest something of the complexity of our relationship with it.

The nearly equal attention given in these essays to television and to the approaches that might be employed in its analysis is in part a result of our desire to introduce readers who might not be familiar with them to some of the approaches that have most influenced contemporary criticism in general. This emphasis on "method" or "theory" also stems from the shared belief that every critical approach carries with it certain basic assumptions about the goals of criticism, the critic's role in that project, the nature of the "thing" criticism hopes to illuminate, and the kind of knowledge that might be produced by the critical act. As you will no doubt discover as you read through these essays, they not only focus on different aspects of television, but they also define television in somewhat different terms. This possibly confusing state of affairs is not a function of the approaches chosen for inclusion in this volume. Everyone who "talks" critically about television does so within a particular theoretical framework. In much of the television criticism you encounter in newspapers and magazines (but also in more "scholarly" contexts as well) the theoretical framework employed is implicit, or the critic maintains that he or she has no particular theoretical "bias." The authors in this volume make their theoretical positions explicit and, in doing so, remind us of the inseparability of the object of criticism from the approach used in its analysis.

Although they would accord it differing degrees of importance, all the contributors to this collection would acknowledge the institutional nature of television. That is to say, commercial television is much more than just a collection of individual programs or "texts" that happen to be interrupted by commercials. It is an institution that exists primarily to translate the phenomenon of simultaneous mass viewing into a commodity that can be sold to advertisers. As Todd Gitlin puts it in his institutional study of commercial television, *Inside Prime Time,* programming executives, "want, above all, to put on the air shows best calculated to accumulate maximum reliable audiences. Maximum dollars attract maximum dollars for advertisers, and advertiser dollars are, after all, the networks' objective. Quality and explicit ideology count for very little." Gitlin goes on to quote a network vice-president as saying, "'I'm not interested in pro-social values. I have only one interest. That's whether people watch the program. That's my definition of good, that's my definition of bad.'"[3]

All works of art and all forms of entertainment are produced and consumed within institutional contexts. The image of the solitary artist pouring his or her soul into poetry that will be read aloud to a select and duly appreciative group of art lovers will hardly suffice as a model for either the production of art or its consumption in the latter part of the twentieth century (if, indeed, it ever did). This romantic image ignores that which is most characteristic of contemporary cultural production: its inescapably institutional and economic nature. Hence, to the scholars represented in this book, commercials, station promotions, and network logos are just as much texts to be analyzed as the programs they surround.

By now it should be pretty clear that contemporary criticism represents a fundamental departure from what we might call pre-structuralist or traditional criticism: the generally accepted set of assumptions about literature and the critical act that has governed literary criticism in the United States for all but the last twenty years or so and continues to condition what we common-sensically accept "literature" and "criticism" to be. Whereas traditional criticism conceives of its object of study as a unified "work," contemporary criticism takes as its object of study the "text": the site of intersection for a complex web of codes and conventions. Whereas traditional criticism emphasizes the autonomy of the artwork, contemporary criticism foregrounds the relationships between texts and the conventions underlying specific textual practices. Traditional criticism is artist-centered; contemporary criticism foregrounds the contexts within which authorship occurs and the forces that circumscribe it. Traditional criticism looks to great art to reveal enduring truths about the world; contemporary criticism considers the worlds constructed within texts. Traditional criticism conceives of meaning as a property of an artwork; contemporary criticism views meaning as the product of the engagement of a text by a reader or by groups of readers. Traditional criticism frequently sees as its function not only the establishment of what a work means but also the separation of "literature" from "non-literature" and the erection of a hierarchy of greatness among works. Contemporary criticism examines the criteria by which those in a position to define literature make such determinations and would expand the scope of literary studies to include both "non-literature" and critical discourse about texts.

It should also be clear why commercial television cannot easily be accommodated within the assumptions of traditional criticism. Commercial television refuses to be broken down into a series of autonomous works, appears authorless, shares few of the qualities we generally associate with great art, makes few demands on the viewer, and would thus seem to leave little for the critic to interpret. Furthermore, commercial television seldom makes any pretense of being "art"; the regular and insistent interruption of programs by commercials reminds us that programs are really only "pre-texts" for the real content of television—advertising messages. The uneasy fit between commercial television and assumptions of traditional criticism partially explains the relative lack of a tradition of television criticism in the United States. It also helps to account for the fact that, in this country, at least, the "golden age" of traditional television criticism corresponds with the "golden age" of television—that brief period of live, original television drama in the 1950s. Such self-contained, "serious" television dramas as *Marty, Requiem for a Heavyweight, Visit to a Small Planet, The Rack,* and *The Death of Billy the Kid* most closely resembled the model of dramatic art with which traditional critics felt comfortable.

The assumptions of contemporary criticism also set it apart from the research tradition that has informed mass communication studies in the United States for the better part of fifty years. Perhaps because broadcasting (at least since the late 1920s) has been thought of more as an advertising vehicle than an artform, research into the relationship between broadcasting and its audience has been largely sociological, rather than aesthetic, in orientation. The emergence of radio as a national advertising medium around 1930 spurred the growth of basic demographic and marketing research in broadcasting. Broadcasters and advertisers needed to know who was listening to what stations, at what times, and with what frequency.

Some attention was paid to what we might in retrospect see as the more "aesthetic" dimensions of radio listening—what types of programs appealed to what types of listeners and why?—but these questions tended to be asked within a social survey framework: "Which programs do you most enjoy listening to?" Where answers were proposed to the question "Why do certain audiences enjoy certain types of programs?" they were generally sociological in nature. For example, after conducting a content analysis of forty-three radio soap operas in 1941, Rudolph Arnheim concluded that they "attract the listener by offering her a portrait of her own shortcomings, which lead to constant trouble, and of her inability to help herself. In spite of the unpleasantness of this picture, resonance can be enjoyed because identification is drawn away from it and transferred to an ideal type of the perfect, efficient woman who possesses power and prestige and who has to suffer not by her own fault but by the fault of others."[4]

The advertising basis of broadcasting also prompted studies of the effects of broadcast messages on the attitudes and behavior of radio listeners. If radio commercials could not affect the decision to purchase a particular product, then, obviously, millions of advertising dollars were being wasted. The perceived need for effects research was heightened during World War II, especially in light of the use of radio as an information source and propaganda vehicle. Numerous studies investigated the possibility that, effectively utilized and tightly controlled, radio might shape the public opinion of an entire nation—for good or for evil. With the rapid growth of television as a popular entertainment medium in the early 1950s, research interests shifted to its effects on viewers and the functions served by television viewing for various subaudiences. Of particular concern was the effect of television viewing on the attitudes and behavior of children.

Over the past thirty years, sociological inquiry into the relationship between television viewers and programs has broadened in scope to include television's possible effects on the conduct of political campaigns; the behavioral and attitudinal consequences of viewing television violence; and television's depiction of minorities, women, and other segments of society; among many other topics. Although some researchers continue to attempt to measure the direct effects of viewing particular programs on particular audience groups, the "hypodermic" or direct effects model of media research has largely given way to models of media-audience interaction that emphasize the functions served by media use and the longer-term and more subtle consequences of media consumption.[5]

Since its emergence as a recognizable field of study in the 1930s, mass communication research in the U.S. has turned to the natural and physical sciences for its model of how knowledge about media-audience relationships might be generated. The application of the "scientific method" to media research is a result, in part, of the need of pioneering research administrators such as Paul Lazarsfeld to legitimize media research and, as a result, earn for it a place within the academic community. Broadcasting organizations, which funded much of the early research on media and media audiences, required that the findings that emerged from these studies be "objective" and "scientific," rather than merely the expression of the investigator's opinion. Thus media research methods were made to resemble those of the physical science laboratory wherever possible. Safeguards were established to minimize the possible effects of the investigator's own expectations upon the results of the study, investigatory procedures attempted to reduce the phenomenon being studied to a limited set of variables (preferably an "independent" and a "dependent" variable), and results were expressed in quantitative terms.

The usefulness of the above model for explaining the complex nature of our relationship with television increasingly has been challenged over the past few years. Some scholars have charged that the application of research procedures from chemistry or biology to social and aesthetic phenomena is inappropriate; they argue that such phenomena cannot be reduced to the investigatory simplicity of the laboratory experiment. Furthermore, they claim that the scientist's belief in the objectivity and value-free status of his or her undertaking is illusory. Obviously, other scholars would refute these objections, while still others would acknowledge the limitations of quantitative research methods and statistical data analysis in explaining certain aspects of media-audience interaction.[6]

Regardless of how successful one believes mass-communication research has been in explaining our relationship with television, however, it should be apparent that a model of media research based upon statistics and quantitative research methods and guided by the goals of laboratory science leaves little room for "criticism" as it is generally conceived in other disciplines. The closest that one can come to "scientific" criticism would be what is called quantitative content analysis, in which one reduces the text (a television program, for example) to quantifiable data by noting the incidence of certain features and comparing that frequency with something else. For example, a researcher might count the number of black characters in a particular program in order to relate their proportion of the program's "population" to that in other programs, in other genres of television programming, or in the general population.[7]

Content analysis can be quite helpful in documenting the ways in which television programs, both fictional and nonfictional, represent constructions of the world rather than reflections of reality. For example, it can help to demonstrate television news's predilection for certain types of stories (natural disasters and political violence, for example) and its disinclination to cover other matters (complex economic problems and the political and social contexts of events in the third world). However, content analysis's adherence to the principles of laboratory science greatly limits its ability to account for the complexities and subtleties of our engagements with those constructed worlds we see on television. We don't experience a situation comedy or a soap opera as a set of quantifiable and "objective" categories of data. When a researcher reduces those types of texts to those types of data it is debatable whether he or she is really studying a sitcom or a soap opera at all—such is the difference between the text the viewer experiences and the data categories into which the text is transformed by the analyst.

Thus, some mass-communication researchers might regard the following essays as representing "subjective" or "unscientific" approaches to media study, because their procedures do not derive from the model of laboratory science and their interpretative strategies are not based on the quantitative analysis of data. To them we would say that the complex relationship between viewer and television may well be so complex and multidimensional that it resists all attempts to reduce it to a phenomenon that can be explained by the procedures that seem to work so well for the chemist. What scientific law explains the curious relationship we have with fictional television programs, for example? We know that the characters and situations presented to us are not "real," that a character who "dies" on a soap opera is played by an actor who will go home at the end of the day and have dinner just like always. And yet those characters and situations are endowed with sufficient "real-seemingness" that they can move us to anger or to tears. How can reducing the world of the soap opera to a set of content categories account for this paradox? What will the observable behavior of viewers or their responses to a survey questionaire tell us about our willingness to "suspend our disbelief" every time we enter the narrative worlds of *Dallas* or *General Hospital*?

This is certainly not to say that there is something inherently wrongheaded about the use of quantitative methods or statistical procedures in mass-communication research. Nor is it to argue that the alternative to quantitative research is a flight into impressionistic opinions about television to which no standards of rigor or validity can be applied. Rather, it is to point out that there are theories and approaches developed largely in other disciplines (literature, film, cultural studies), and informed by a different set of philosophical assumptions from those that underlie traditional American media research, that might provide fresh insight into our relationship with television.

These essays are not arranged in order of perceived significance (either ascending or descending) or in the belief that the final chapters provide the ultimate answers to the problems left unsolved in earlier chapters. We begin with Ellen Seiter's essay on the relationship between semiotics and television. Seiter reviews the basic vocabulary and fundamental tenets of structural linguistics and semiotics. The essay's placement at the beginning of the collection acknowledges the fact that the terminology and approach of semiotics inform all of the other essays in the collection. Saussure provided us with the insight (among many others) that meaning is produced in language not through any direct connection between words and things, but as a result of an endless circuit of reference among words. Working independently, but at roughly the same time, Peirce extended this fundamental precept of modern linguistics to nonlinguistic communication. Even visual language systems (photography, cinema, and television) operate by connecting images with culturally conditioned concepts and not by automatically reflecting the immanent meaning of things. As Seiter puts it, "Peirce's concept of the sign is important because it forces that realization that no communication takes place outside of sign-systems: we are always translating signs into other signs." Semiotics gives us a vocabulary with which to describe the relationships among signs on television. Seiter's analysis of the opening credit sequence of *The Cosby Show* uses that terminology to demonstrate the semiotic complexity of a brief and apparently straightforward moment of television—a moment that millions of us have watched dozens of times.

With Sarah Kozloff's essay we move from a general theory of the nature of and relationships among all types of signs to a consideration of a specific kind of semiotic organization—the narrative. A discussion of narrative theory is crucial to television criticism because, as Kozloff puts it, "[t]elevision is the principal storyteller in contemporary American society." Furthermore, as she also demonstrates, so much of what we experience on television is structured as narrative—even types of television programming we don't usually think of as stories (a *60 Minutes* segment on San Diego, for example). Narrative theory attempts to account for patterns of organization that run through all narratives and for differences among narratives. Kozloff makes the case that television series maintain our interest by giving us predictable stories that are combined in novel ways. She also discusses the particularities of television as a vehicle for storytelling: the embedding of individual stories within a continuous flow of discourse, the inherent "interruptability" of television narratives, and the attenuation of narratives over years of programming in the series and serial forms.

My own essay takes up the flipside of narrative theory. Narrative theory addresses the fact that every narrative is not just a story, but a story told by someone (or, in the case of television, by some "thing") and in a particular way. Reader-oriented criticism begins with an equally common-sensical insight: every story is told to someone and is made sense of by the "listener" in particular ways. Reader-oriented criticism attempts to describe what happens when we read a fictional narrative and to describe the places marked out for hypothetical readers within texts (what Kozloff discusses briefly in the previous chapter as the "narratee"). I argue that this type of criticism can help to explain our curious relationship with the worlds of daytime soap operas— a relationship involving a "reading act" that can be stretched out, quite literally, over decades. I also use reader-oriented criticism to examine the two principal modes of viewer engagement in television: the "Hollywood" mode, by which we are engaged covertly; and the "rhetorical" mode, by which we are addressed directly.

In her essay on genre theory and television, Jane Feuer discusses another important aspect of our relationships with texts in general and television texts specifically; she argues that part of the process of making sense of any given text involves identifying that text as part of a larger body of texts with which it shares certain features. Genre theory helps to provide us with ways of relating industrial practice in television (the need to turn out on a regular schedule huge numbers of texts that are likely to appeal to millions of different viewers) to the texts that are produced as a result of this practice, and both to the expectations of audiences. Television genres (the sitcom, the crime drama, the prime-time soap) represent the industry's attempt to predict audience popularity, whereas, as Feuer puts it, "[f]or the audience . . . genre assures the interpretability of the text." Feuer considers several different conceptualizations of genre and relates them to one of the most enduring of television forms, the sitcom.

Mimi White's essay on ideological analysis reminds us that our relationship with television is never "innocent," regardless of how innocuous its stories might seem. Rather, by agreeing to be viewers, we implicitly become parties to a contract between ourselves and an enormous institution that makes the pleasure we derive from watching television the basis for our transformation into commodities. In television it is the viewer, not the product advertised, that is literally "sold." Ideological analysis concerns itself with the nature and functioning of television as institution, the assumptions and values that underlie the texts it produces, and the manner by which we are positioned in relation to both institution and text. As White demonstrates, even the act of watching a rather unremarkable commercial carries with it an enormous range of assumptions about television in general and the cultural contract we make with it.

Sandy Flitterman-Lewis takes up a relationship that has already been examined to some degree in each of the previous chapters—that curious relationship between the mind of the viewer and the world of the television text. She reconceptualizes this relationship within the framework provided by contemporary psychoanalytic theory, especially as it has been applied to film spectatorship. Flitterman-Lewis takes as her starting point the suggestive analogy between the act of dreaming and that of watching a film—an analogy at the heart of psychoanalytic film theory. Although acknowledging the relevance of that analogy in connecting the desires of the spectator with the fantasies enacted in visual narratives, Flitterman-Lewis goes on to demonstrate the important differences between film and television from a psychoanalytic perspective and the resultant modifications that must be made in cine-psychoanalysis if it is to serve as the basis for understanding the relationship between television text and "tele-spectator."

Throughout the history of American broadcasting, the industry has regarded women as the prime audience for most types of programming. Indeed, since the early 1930s, programming during daylight hours has been designed almost exclusively to attract the largest possible number of women between the ages of eighteen and forty-nine—the portion of the population that makes the vast majority of consumer purchasing decisions. In her essay on feminist criticism and television, E. Ann Kaplan considers both how women are represented on television and how women as television spectators are engaged by the medium. Kaplan surveys the variety of feminist approaches to television that have been developed over the past twenty years, giving particular attention to what she terms "post-structuralist" feminism: the cluster of approaches that grow out of an analysis of "the language order through which we learn to be what our culture calls 'women' . . . as we attempt to bring about change beneficial to women." Poststructuralist feminism has appropriated and modified some of the insights of psychoanalysis, particularly its account of the manner by which we are given positions within the family and society according to gender. Because this type of feminist analysis of mass media has been elaborated principally within film theory, Kaplan considers both what feminist television scholars might borrow from film theory and what must be reconsidered in light of the differences between film and television. In a case study of the application of feminist criticism to television, Kaplan focuses on music videos and their arrangement as a continuous flow of programming on MTV. She asks: "Is there any particular gender address in this flow? . . . [And] how does this kind of flow particularly affect the female spectator?"

In the final essay, John Fiske examines the strand of television analysis that has emerged from contemporary British cultural studies, especially the work of scholars connected with the Birmingham University Centre for Contemporary Cultural Studies. As Fiske notes, the discipline of British cultural studies is not so much a single critical approach as it is an attempt to work out relationships among a number of separate approaches—ideological analysis, semiotics, psychoanalysis, feminist criticism, and ethnographic anthropology. The British cultural studies tradition relies on a Marxist conception of culture, and in the first part of his essay, Fiske discusses Louis Althusser's conception of ideology and Antonio Gramsci's notion of hegemony (continuing a discussion begun by Mimi White). Within the overall framework of ideological analysis, British cultural studies has conceived of culture as an arena of struggle between those with and those without power. Watching television, then, is not a process by which ideological messages are implanted in the consciousness of a uniform mass audience, but rather a process of negotiation between groups of viewers in different social situations and television texts that are themselves open to a variety of interpretations. Of all the approaches covered in the book, British cultural studies places the greatest emphasis on how actual viewers make use of television in their own lives. Thus in his analysis of Madonna as a popular culture phenomenon, Fiske considers not only the structure of Madonna's music-video texts but also responses to those texts (and to other manifestations of the Madonna image) by her fans.

Each chapter concludes with suggestions for further reading on the critical approach dealt with and includes key works of television criticism produced from that approach. A more general bibliography of television criticism is provided at the end of the book.

The reader should keep in mind that these essays in no way do justice to the complexities of the individual strands of contemporary criticism they consider. Ours is not an attempt to give a *Reader's Digest* version of narratology or psychoanalysis as a substitute for grappling with the central works in those fields. Rather, we lay out in a provisional and necessarily schematic fashion some of the ways these approaches might aid in an understanding of television. We leave it to the reader to explore these critical frameworks further and to test and challenge the relationships between contemporary criticism and television that we have proposed.

NOTES

1. Jonathan Culler, *The Pursuit of Signs: Semiotics, Literature, and Deconstruction* (London: Routledge and Kegan Paul, 1981), p. 25.

2. Pierre Macherey, *A Theory of Literary Production*, trans. Geoffrey Wall (London: Routledge and Kegan Paul, 1978), p. 48.

3. Todd Gitlin, *Inside Prime Time* (New York: Pantheon, 1985), pp. 25, 31.

4. Rudolph Arnheim, "World of the Daytime Serial," in *Radio Research, 1942–43*, ed. Paul Lazarsfeld and Frank Stanton (New York: Duell, Sloan, and Pearce, 1944), p. 60.

5. The shift from a "high effects" to a "low effects" model of media is discussed by Todd Gitlin in "Media Sociology: The Dominant Paradigm," *Theory and Society* 6 (1978): 205–53. For examples of the functionalist approach to media study, see Jay G. Blumler and Elihu Katz, eds., *The Uses of Mass Communication* (Beverly Hills, Calif.: Sage, 1974).

6. A good overview of the philosophical and methodological debates currently being conducted in communication research is provided by the special "Ferment in the Field" issue of *Journal of Communication* 33 (Summer 1983). My own objections to "empiricist" mass-communication research are laid out in the first two chapters of *Speaking of Soap Operas* (Chapel Hill: University of North Carolina Press, 1985).

7. For examples of quantitative content analysis, see Mary Cassata and Thomas Skill, eds., *Life on Daytime Television: Tuning in American Serial Drama* (Norwood, N.J.: Ablex, 1983); and Bradley S. Greenberg, ed., *Life on Television: Content Analysis of U.S. TV Drama* (Norwood, N.J.: Ablex, 1980).

SEMIOTICS AND TELEVISION
ELLEN SEITER

Semiotics is the study of everything that can be used for communication: words, images, traffic signs, flowers, music, medical symptoms, and much more. Semiotics studies the way such *signs* communicate and the rules that govern their use. As a tool for the study of culture, semiotics represents a radical break from traditional criticism, where the first order of business is interpretation of an aesthetic object or text in terms of its immanent meaning. Semiotics first asks *how* meaning is created, rather than *what* the meaning is. In order to do this, semiotics invented a specialized vocabulary to describe signs and how they function. Often this vocabulary smacks of scientism to the newcomer and clashes with our assumptions about what criticism and the humanities are. But the special terminology of semiotics and its attempt to compare the production of meaning in a diverse set of mediums—aesthetic signs being only one of many objects of study—have allowed us to describe the workings of cultural communication with greater accuracy and enlarged our recognition of the conventions that characterize our culture.

The term semiotics was coined by Charles S. Peirce (1839–1914), an American philosopher interested in pragmatism; his work on semiotics did not become widely known until the 1930s. The field was also "invented" by a Swiss linguist named Ferdinand de Saussure. The term he used to describe the new science he advocated in *Course in General Linguistics,* published posthumously in 1959, was semiology. Beginning in the 1960s, some leading European intellectuals devoted themselves to the study of semiotics and applied it to many different sign systems. Roland Barthes carefully analyzed fashion, French popular culture, and a novella by Balzac. Umberto Eco turned his attention to Superman comic strips and James Bond novels. (In 1983, Eco published an international best seller, a peculiarly semiotic detective novel entitled *The Name of the Rose.*) Christian Metz set out to describe the Hollywood cinema as a semiotic system. At its best, semiotics supersedes some of the apparently arbitrary divisions between fields of study that we are familiar with as academic departments in the university; it draws ideas from many different disciplines, and this makes it specially suited to the study of television.

THE SIGN

The smallest unit of meaning in semiotics is called the sign. Semiotics begins with this smallest unit and builds to rules for the combination of signs and the connotative meanings produced from them. Ferdinand de Saussure conceptualized the sign as composed of two distinct parts, although these parts are not separable in actual communication, only in theory. Every sign is composed of a *signifier,* that is, the image, object, or sound itself—the part of the sign which has a material form— and the *signified,* that is, the concept it represents. In written language the sign *rain* is composed of the four letters on this page (the signifier) and the definition of rain (the signified), that is, drops of water that fall from the sky. Saussure argued that the relationship between the signifier and the signified in language was entirely conventional, completely arbitrary. Words have no positive value, no meaning in and of themselves. Instead, a word's meaning derives from its difference from other words in the sign system of language. On the level of signifier, *rain* derives its meaning from its distinguishability from brain or sprain or rail or Braille or roan or reign. The signified is meaningful because of its difference from sprinkle, drizzle, downpour, monsoon. Each language marks off a set of meaningful differences: we can imagine an infinite number of possibilities for signifiers and signifieds, but each language makes only some differences important and detectable. Learning a second language is difficult because each language consists of a set of signs whose meanings derive from differences we might not be sensitive to, such as phonetic distinctions, grammar rules, and words that are untranslatable into our first language.

Saussure's ideas are more difficult to apply to sign systems like television, in which language is only one component. When the word *rain* is used on a television weather forecast, for example, we are faced with a much more complex signification. The signifier is made up of many different components of the image and the soundtrack. The two-dimensional, luminous image is rectangular, with a fixed aspect ratio of 3 : 4. The focal length of the camera lens, lighting, angle, color, and composition are characteristics of the signifier as well. The signified can be thought of as everything that this television image and sound-track represent: the newscaster, his voice, speech pattern, dress, hair style, make-up, as well as the meaning of any words that may appear on the screen, such as the weather forecast. When watching television, we often forget about the signifier and treat television's signs as pure signified.

Semiotics allows us to recognize the conventional and arbitrary relationship of signifier to signified in a whole range of signs that we take for granted in our everyday life as natural, even necessary. It also allows us to describe the relationships among signs within a single system, such as television. Umberto Eco defines the sign this way: "A sign is everything that, on the grounds of a previously established social convention, can be taken as something standing for something else."[1] Eco means to include in this definition natural signs as well as cultural ones: it is through social convention that we learn to interpret a dark, cloudy sky as signifying the approach of a rainstorm.

Eco's conception of the sign is adapted from the work of Charles S. Peirce, who did not limit himself to linguistics, as Saussure did, but attempted to account for every type of sign. For Peirce, the sign could be broken down into three parts, rather than Saussure's two. These were the representamen (roughly equivalent to the signifier), the object (or signified) and the interpretant (the sign that we use to translate the first sign). The interpretant is what we use to describe the sign, and in our description we inevitably turn to another sign. When we say that an image on the television news is "Winnie Mandela," we have translated the sign system of television images into another sign system—that of proper nouns. This interpretant is another sign that translates, explains, stands for the first sign. Peirce saw this as an unending chain of sign production, what Eco has dubbed "unlimited semiosis." The sign "Winnie Mandela" may require a further interpretant for some, such as "South African leader in the fight against apartheid." This may lead to more signs if we attempt to describe "Winnie Mandela": more words, more images, and so on.

Peirce's model does not require the *intention* to communicate: signs may be produced by nonhuman agencies, for example, or by unconscious senders. Peirce's model does not necessarily require a human receiver of the sign, or any receiver at all, though, as Eco explains, "the interpretation by an interpreter, which would seem to characterize a sign, must be understood as the possible interpretation by a possible interpreter. . . . The human addressee is the methodological (if not the empirical) guarantee of the existence of signification."[2]

A mechanical failure at my local television station may send the sign "technical breakdown" to my television set, whether I am in the room watching it or not; in this case there is a nonhuman sender and no receiver at all. Peirce's concept of the sign is important because it forces the realization that no communication takes place outside of sign systems—we are always translating signs into other signs. The conventions of the sign system control the ways we are able to make meanings (that is, produce signifiers) and limit the range of meanings available to us (that is, what signifieds we produce).

Peirce described signs as falling into three categories: symbolic, iconic, or indexical. Language uses symbolic signs, in which the relationship between signifier and signified is an arbitrary one. But language can fool us into feeling as though the sign is somehow motivated by a natural connection to what it stands for. When we study a second language, the discovery of "cognates" (words which remain substantially the same from one language to another) may reinforce our feeling of the necessary associations between certain words (or signifiers) and certain signifieds. But many words bear no similarity whatsoever to our first language, even if they are from the same language family; and in each language some things are said in ways that are impossible in other sign systems. Even onomatopoeia—the use of words that seem to imitate the sound they signify—turns out to be conventional after all. For speakers of English the rooster goes "cock-a-doodle-doo"; for Germans he goes "Kikeriki."

Language is a very important sign system used by television, but words are not the only symbolic signs that television uses. Objects may become symbolic signs, as when roses signify love; champagne signifies celebration; a rainbow symbolizes hope; a flag symbolizes a nation. Symbolic signs may be invested with so much feeling that the conventionality of the connection between the signifier (for example, a cloth patterned in red, white and blue) and the signified (United States of America) is no longer easy to grasp. The mistreatment of certain signifiers, such as a cross, may be understood as a direct attack on the signified (Christianity). Other symbolic signs, such as red lights, may be invested with less emotion, but may also feel more or less "natural" to us: we *feel* as though red and only red could be used to tell us to stop.

Television incorporates many symbolic signs into its images; we then have a sign (the image) that stands for the symbolic sign (the object). Some aspects of the image that we think of as nonrepresentational function as symbolic signs as well, such as colors (pink for femaleness, white for goodness); music (minor chords and slow tempos to signify melancholy, solo instrumentals to signify loneliness); photographic technique (soft focus to signify romance, hand-held camera to signify on-the-spot documentary). These signs are all established through convention, through repeated use. Semiotic analysis of television has as one of its goals making us conscious of the use of symbolic signs on television, so that we realize how much of what appears "naturally" meaningful on TV is actually historical and changeable.

The second kind of sign which Peirce described is the iconic sign, in which the signifier structurally resembles its signified. We must *learn* to recognize this resemblance, as when we learn to read maps or to draw. The correspondence between a drawing of a cat, for example, and the signified "cat" (which might be a particular specimen of cat or the concept of cat in general) could take many different forms. The drawing could be skeletal or anatomical, in which case it might take the training of a veterinarian or a zoologist to recognize any structural similarity between the drawing and the signified "cat." The iconic sign could be a child's drawing, in which another kind of expert decoder, such as the child's parent or teacher, might be required to detect the structural resemblance. Most drawings rely on rules that dictate point of view and scale; an "aerial view" of a cat, a head-on angle, or a drawing done twenty times scale would be much harder for most of us to recognize than the conventional side-angle view where two legs, a tail, a pointed ear and whiskers will do the job, even if no attempt is made to render coloration and the entire drawing appears only as an outline in black. Most of these admonitions about the conventionality of drawings hold true for video images as well, even though we think of television as much more lifelike. By violating conventions of scale, perspective, camera angle, color, lighting, lens focal length, and subject-to-camera distance of focus, we could easily obtain a video image of a cat that would defy recognition—no one would guess that the object in front of the camera had been a cat. Even iconic signs that we treat as particularly unique because they have as their signified an individual

living creature may be dictated by convention. The feline celebrity
Morris, from the Nine Lives cat food commercials, died, but his fans
did not have to suffer the loss of his image. The iconic sign "Morris"
lives on, thanks to the skills of the production crew and the animal
trainer who found a stand-in cat to make the commercials. It may be a
blow to our faith in physiognomy, but we can be fooled by pictures of
persons almost as easily.

The structural resemblances involved in iconic signs must be
learned, and with TV images this involves learning to recognize many
conventions of representation. One of the characteristics of such representational codes is that we tend not to recognize their use; they become as "natural" to us as the symbolic signs of language, and we
think of iconic signs as the most logical—sometimes as the only possible—way to signify aspects of our world.

Indexical signs, the third type of sign defined by Peirce, involve an
existential link between signifier and signified: the sign relies on their
co-presence at some point in time. Drawings do not qualify as indexical signs because we can make a drawing of something we have
never seen. Maps are iconic, rather than indexical, because a car-
tographer can create a map solely on the basis of other iconic signs,
such as diagrams and geological surveys; she may never have been to
the place the map will "signify." Indexical signs are different from
iconic ones because they rely on a material connection between signifier and signified: smoke means fire; pawprints mean the presence
of a cat; a particular set of fingerprints signifies "Richard Nixon"; red
spots signify "measles." Most images produced by cameras belong to
the class of "indexical signs" because they require the physical presence of the signified before the camera lens at some point in time
for their production. This fact about an image is difficult to verify,
however, and images—like those of Morris—can lie. Many images
produced by cameras do not meet such qualifications, such as trick
photographs, special effects, computer-generated graphics, multiple
exposures, and animated images.

Indexical signs are established through social convention. Animals
have left pawprints as long as they have roamed the earth, but their
pawprints became a sign only when people began to use them for
tracking. As Umberto Eco explains:

> The first doctor who discovered a sort of constant relationship be-
> tween an array of red spots on the patient's face and a given dis-

ease (measles) made an inference: but insofar as this relationship has been made conventional and has been registered as such in medical treatises a *semiotic convention* has been established. There is a sign every time a human group decides to use and recognize something as the vehicle for something else.[3]

Indexical signs are no less tainted by human intervention than symbolic or iconic ones; they require the same accumulation of convention, the same reinforcement and perpetuation within a society to be understood as signs in the first place. These categories are not mutually exclusive. Television constantly uses all three types of signs, and after one has understood the distinctions among them it may be useful to think of television functioning as two or three different kinds of signs simultaneously—for example, as symbolic *and* indexical.

"The camera never lies" is a statement that tells us a lot about the way we accept television images as real because they involve indexical signs, even if, from a semiotic point of view, the statement is a falsehood. Many television images are produced in such a way that we are encouraged to understand them only as indexical signs. Stand-up shots of reporters on location is one example of this; we may not be able to decipher from the image itself whether Andrea Mitchell is really standing on the White House lawn or not, but TV places an enormous stress on the *connection* between the image and this location as it exists in real time and space.[4] So much has been made of the objectivity of the camera as a recording instrument since its invention that we often fail to recognize the extent to which camera images are produced according to rules just as drawings are. Semiotics reminds us that the signifiers produced by TV are related to the signifieds by convention, even if, when we watch something like the news, we tend not to think of the active production of signs involved in TV, but receive the news rather as pure information, pure signified.

It is important to remember that TV is a sign system, and, therefore, can be used to lie. To engage in fantasy for a moment, consider producing a newsbreak about an event of pure imagination for broadcast on network TV. If we gave some careful thought to the way newsbreaks are written and the topics usually covered in them, we could script and storyboard a newsbreak on our own. If we had access to the facilities, technicians, equipment, supplies, and personnel of one of the networks, and if we could coerce an anchor to violate his professional ethics (or find a convincing impostor) to read our script, we could produce a newsbreak, complete with "live action" reports, that would be indistinguishable from the authentic item. No viewer could detect the difference from the TV sign alone.

Umberto Eco criticizes Peirce's distinction among symbolic, iconic, and indexical because it tends to overlook the historical and social production of all signs. Instead, Eco offers a definition that casts all signs in terms of this context: "Semiotics is in principle the discipline studying everything which can be used in order to lie. If something cannot be used to tell a lie, conversely it cannot be used to tell the truth: it cannot in fact be used 'to tell' at all."[5]

COMBINATION AND CONNOTATION

A semiotics of television provides us with a set of problems different from those encountered when studying written or spoken language. What is television's smallest unit of meaning? Is there a grammar of TV, a set of rules governing the combination of sounds and images? Where does signification begin and end? How does television as a sign system change? To answer these questions it will be necessary to introduce several more terms from the special vocabulary of semiotics: *channel, code, paradigmatic, syntagmatic, connotation, denotation, synchronic, diachronic*. The late Paddy Whannel used to joke, "Semiotics tells us things we already know in a language we will never understand." Learning the vocabulary of semiotics is certainly one of its most trying aspects. This vocabulary makes it possible, however, to identify and describe what makes TV distinctive as a communication medium, as well as how it relies on other sign systems to communicate. Both questions are vital to the practice of television criticism.

Unlike language, television does not conveniently break down into phonemes, morphemes, or letters of the alphabet. A technological definition of the smallest unit of meaning is the frame: "A complete scanning cycle of the electron beam, which occurs every 1/3 second. It represents the smallest complete television picture unit."[6] If we use the frame as the "smallest unit of meaning," however, we ignore the soundtrack, where 1/30 second would not necessarily capture a meaningful sound. When Christian Metz wrote his semiotics of the cinema he identified five channels of communication: image, written language, voice, music, and sound effects.[7] In borrowing these categories, I substitute the term graphics for written materials, so as to include the logos, borders, frames, diagrams and computer-animated images that appear so often on our television screen, and because this is the production term for such images. In *Cinema and Language,* Metz concluded that television and cinema were "two neighboring language systems" that were characterized by an unusual degree of closeness, but he never analyzed television in the same meticulous way he did the cinema. Before returning to the question of TV's smallest unit of meaning, it will be useful to review some recent theoretical work on how TV uses these five channels and how this usage compares to that of the cinema.

It is a commonplace that TV is nothing but talking heads—which tells us that facial close ups and speech are singularly important to it. Television production textbooks warn students about the need for simplicity in the image and explain how to achieve it through visual *codes* such as symmetrical compositions, color compatibility, and high key lighting. These conventions of TV production represent an *interpretation* of video technology and its limitations, but they are not a necessary consequence of it. As John Ellis has explained the logic of these visual codes: "Being small, low definition, subject to attention that will not be sustained, the TV image becomes jealous of its meaning. It is unwilling to waste it on details and inessentials."[8] In part, these codes dictate both how the images are produced and what is represented: we see more shots of actors, emcees, newscasters, politicians—and commodities—than anything else.

Broadcast TV uses graphics to clarify the meaning of its images, and it does so to a much greater extent than the motion picture. Diagrams are superimposed over news or sports images to invite a quasi-scientific scrutiny of the image. Borders and frames mask out the background of the already pared-down images. Words constantly appear on the screen to identify the program, the sponsoring corporation, the network, the product name, the person portrayed. Often the words on screen echo speech on the soundtrack, which goes on continually. In his analysis of other forms of mass communication, Roland Barthes described verbal language as always providing the definitive meaning for the image: "It is not very accurate to talk of a civilization of the image—we are still, and more than ever, a civilization of writing, writing and speech continuing to be the full terms of the informational structure."[9] For Barthes, verbal language is used to close down the number of possible meanings the image might have. This "anchoring" of the image by the verbal text frequently supplies a bourgeois world-view: "The anchorage may be ideological and indeed this is its principal function; the text directs the reader through the signifieds of the image, causing him to avoid some and receive others; by means of an often subtle *dispatching* it remote-controls him towards a meaning chosen in advance."[10]

John Ellis and Rick Altman have argued that the television sound-track—speech, music, sound effects—entirely dominates the image by determining when we actually look at the screen. The soundtrack is so full, so unambiguous that we can understand television just by listening to it. Because television is a domestic appliance that we tend to have on while doing other things—cooking, eating, talking, caring for children, cleaning—our relationship to the television set is often that of auditor rather than viewer. Altman argues that sounds such as applause, program theme music, and the speech of announcers tend to precede the image they refer to and serve primarily to call the viewer back to the screen: "The sound serves a value-laden editing function, identifying better than the image itself the parts of the image that are sufficiently spectacular to merit closer attention by the intermittent viewer."[11] For Altman, the television soundtrack acts as a lure, continually calling to us: "Hey, you, come out of the kitchen and watch this!"

The television commercial, the football game, the made-for-TV movie, the talk show, and the situation comedy have their own distinctive ways of combining sounds and images. From a semiotic viewpoint, one of the most important characteristics of television in general (and one that is shared by many genres) might be its tendency to use all five channels simultaneously, as television commercials typically do. This might also explain television's low status as an aesthetic text; too much goes on at once on TV and there is too much redundancy among the different sound elements and the image for it to be "artistic." The primary function of the soundtrack violates conventional notions in cinema aesthetics about the necessity of subordinating sound to image. The high degree of repetition that exists between soundtrack and image track, and between segments, is mirrored at the generic level of the series, which is television's definitive form. As Umberto Eco explains about the debased aesthetic status of TV: "This excess of pleasurability, repetition, lack of innovation, was felt as a commercial trick (the product had to meet the expectations of its audience), not as the provocative proposal of a new (and difficult to accept) world vision. The products of mass media were equated with the products of industry insofar as they were produced *in series,* and the 'serial' production was considered as alien to the artistic invention."[12] Because semiotics recognizes the role of convention in all verbal and visual sign production—including aesthetic production—it tends to take a less condemning view of television and therefore may have more to say about TV as a communication system than traditional criticism, which dismisses TV as a vulgarity.

Film semiotics has tended to scrutinize the image much more than the soundtrack. In television this separation is much more problematic, because sound bears an entirely different relationship to the image. To speak of the smallest unit of meaning on a soundtrack we must separate its three components—music, speech, sound effects. Each one has sound elements whose miminum unit is of different duration, and these do not correspond to single images, but overlap the images. A semiotics of TV sound calls for linguistics, audiology, and musical theory and is outside of the scope of this chapter, where we will have to confine discussion to the image.

Christian Metz concluded that the cinema is so different from language that we must be wary of applying linguistic theory. For Metz, no smallest units were discernible in the cinema. It must be analyzed at the level of the shot, which he called its "largest minimum segment." This resembles Eco's conclusion that iconic signs—images—are not reducible to smaller units; they are already "texts," that is, combinations of signs, and they are governed by a code that is weak compared to the grammar rules that govern language. Weak codes are flexible, changeable, and can produce an unforeseeable number of individual signs.[13]

Metz was able to explain a great deal about editing as a code of the classical Hollywood cinema using the shot as his "minimum segment" and the semiotic concepts *paradigmatic* and *syntagmatic.*[14] A syntagm is an ordering of signs, a rule-governed combination of signs in sequence. A paradigm is a set of signs that are similar in that they may be substituted for one another according to the rules of combination. One paradigmatic category, based on subject-to-camera distance, consists of the class of signs we identify as close-ups. Another paradigm might be "all shots of Bill Cosby." When a syntagmatic chain is produced—for example, a sequence of shots on an episode of *The Cosby Show*—the shots defined in terms of these paradigms are inserted in their proper place in the sequence: long shot of Cosby, medium shot of Cosby, close-up of Cosby, close-up of Phylicia Rashād, close-up of Cosby.

The concepts paradigmatic and syntagmatic may be applied to a higher level of organization than the shot. We could define a paradigm as all television commercials. A syntagmatic chain that selects from this paradigm is: closing credits of *The Cosby Show;* cereal commercial; Armed Forces commercial; continued closing credits of *The Cosby Show;* NBC commercial (preview) for special; NBC commercial for Friday night programs; automobile commercial. Because television is often broadcast twenty-four hours a day, and because it is so discontinuous—combining many different segments of short duration—determining the beginning and end of these "syntagmatic chains" presents special problems for the TV critic. Paradigms are not easily determined either. Does it make sense to analyze an individual episode apart from its place in the entire series? Can we separate the program from the commercial breaks when we write television criticism?

Syntagms and paradigms can be found in relationships between texts, as well as within a single text. A generic paradigm of "TV game show" might include *Wheel of Fortune, Let's Make a Deal, The $64,000 Question, Queen for a Day, What's My Line, Jeopardy,* and *Family Feud.* It would be necessary to describe the basis on which these shows can be grouped together, our criterion for associating them. A syntagmatic arrangement of game shows might be based on their sequence in programming, their place on the TV schedule with morning shows first and evening shows later. Another kind of syntagm might be based on their chronological appearance in the course of TV broadcast history, with an older show such as *Queen for a Day* preceding a more recent one such as *Wheel of Fortune.* Paradigmatic associations are *synchronic:* we group signs together as though they had no history or temporal order. Syntagmatic relationships tend to be *diachronic:* they unfold in time, whether it be a matter of seconds or of years. The meaning of every sign is influenced by syntagmatic and paradigmatic relationships. *Wheel of Fortune* derives some of its meaning from its differences from and similarities to other TV game shows. It also derives meaning from its position on the daily TV schedule, and its place on the time line of broadcast history.

Saussurean linguistics is a synchronic model for the study of language. It insists that sign systems can only be understood as they exist at one point in time. Semiotics was founded, then, on a static model of the sign. Some of the gravest shortcomings of semiotics as a theory are a consequence of this: it inherits the tendency to ignore change, to divorce the sign from its referent, and to exclude the sender and receiver. These characteristics limit the usefulness of semiotics in the study of television. Because television is based on weaker codes than those that govern language, it is much more unstable as a system of meaning. Unlike language, television does not provide most of us with access to sign production. Terry Eagleton's critique of structuralism (which includes semiotics) in the study of literature must be remembered when studying television, as well:

> Structuralism, in a word, was hair-raisingly unhistorical. . . . Having characterized the underlying rule-systems of a literary text, all the structuralist could do was sit back and wonder what to do next. There was no question of relating the work to the realities of which it treated, or to the conditions which produced it, or to the actual readers who studied it, since the founding gesture of structuralism had been to bracket off such realities. In order to reveal the nature of language, Saussure . . . had first of all to repress or forget what it talked about: the referent, or real object which the sign denoted, was put in suspension so that the structure of the sign itself could be better examined.[15]

Semiotics is extremely useful in its attempt to describe precisely how television produces meaning, and in its insistence on the conventionality of all signs. For if signs are conventional they are also changeable. But semiotics remains silent on the question of how to change a sign system. Stubbornly restricting itself to the text, it cannot explain television economics, production, history, or the audience.

The problem with treating television as a synchronic system is best illustrated with the case of connotative meanings. Roland Barthes devoted much of his work to the distinction between *denotation* and *connotation* in aesthetic texts, including images. Denotation is the first order of signification: the signfier is the image itself and the signified is simply what was before the camera, what the signifier is a picture of. Connotation is a second order signifying system that uses the first sign, the denotation, as its signifier, and attaches another meaning to it, another signified. Connotation tends to obscure the status of the picture itself as a sign, that is, the first order of signification. Barthes thought of connotation as fixing, or freezing, the meaning of the denotation; it impoverishes the first sign by ascribing a single and usually ideological signified to it.[16] This is why it takes many words to describe the signifier at the first level: we must include camera angle, color, size, lighting, composition, and so on. But connotations can be described in just a few words.

To begin with a simple denotation: the fade to black has as its signifier the gradual disappearance of the picture to black on the screen, and as its signified simply "black." This sign has been strongly conventionalized in the syntagmatic code of motion pictures and television so that it exists as the following connotative sign: the signifier is "fade to black" and the signified is "the end." Television production texts insist students must always fade to black at the end of every program and before any commercial breaks.[17] The fade to black has become part of a very stable connotation. But connotations have a way of multiplying and changing. On *Knot's Landing,* a CBS prime time soap opera that has cultivated an image as a "quality program," each segment ends with a fade to black that lasts several beats longer than on most programs. Here "fade to black" is part of the tone of *Knot's Landing;* it is used for the connotation "serious drama," "high class show." Connotations fix the meaning of a sign, but the denotation "fade to black" could take on other meanings, as well. A frequent use of fades to black could connote "rank amateur production," for example, or "experimental, modernist video."

Barthes argued that connotation is the primary way that the mass media communicate ideological meanings. A dramatic example of the operation of "myth," as Barthes called such connotations, and of television's rapid elaboration of new meanings is the space shuttle *Challenger.* The sign consisted of a signifier—the TV image itself—that was coded in certain ways (symmetrical composition, long short of shuttle on launching pad, daylight, blue sky background) for instant recognition, and the denoted meaning, or signified, "space shuttle." On the connotative level, the space shuttle was used as a signifier for a set of ideological signifieds such as scientific progress, manifest destiny in space, U.S. superiority over the U.S.S.R. On 28 January 1986, these connotations were radically displaced by the explosion of the *Challenger.* That day all three commercial networks broadcast videotape of the space shuttle repeatedly, obsessively. It was acccompanied first by a stunned silence, then by an abundance of speech by newscasters, by expert interviewees, by press agents, by President Reagan (who canceled his State of the Union address to speak about the explosion), much of which primarily expressed shock. The connotation of the sign "space shuttle" was destabilized; it became once again subject—as a denotation—to an unpredictable number of individual meanings or competing ideological interpretations. It was as if the explosion restored the sign's original signified, which could then lead to a series of questions and interpretations of the space shuttle that related to its status as a material object, its design, what it was made out of, who owned it, who had paid for it, what it was actually going to do on the mission, who had built it, how much control the crew or others at NASA had over it. At such a moment, the potential exists for the production of counterideological connotations. Rather than "scientific progress," the connotation "fallibility of scientific bureaucracy" might have been attached to the space shuttle; "manifest destiny in space" might have been replaced by "waste of human life"; and "U.S. superiority over the U.S.S.R." by "basic human needs sacrificed to technocracy."

Television played a powerful role in stabilizing the meaning of the space shuttle. The networks, following the lead of the White House, fixed on a connotation that was compatible with the state ideology almost immediately. This connotative meaning is readable in the graphic that was devised by the television production staffs and appeared in the frame with the newscasters when they introduced further reports on the Challenger: an image of the space shuttle with a U.S. flag at half-mast in the left foreground. Television fixed the connotation "tragic loss for a noble and patriotic cause" to the sign "space shuttle." Television produced this new connotation within hours of the historical event of the *Challenger*'s explosion. Some of its force comes from its association with cultural and ideological codes that already have wide circulation: the genre of war films, the TV news formula for reporting of military casualties, the history of national heroes and martyrs. Later interpretations of the *Challenger* explosion or the space shuttle program must compete with this one, which is in turn vulnerable to historical change.

The study of connotation indicates the importance of understanding television signs as a diachronic or historical system—one that is subject to change. Semiotics allows us to describe the process of connotation, the relationship of signs within a system, and the nature of signs themselves. But the study of connotation also directs us outside the television text and beyond the discourse of semiotics. We cannot comprehend the space shuttle as a sign without studying the producers of messages (television networks, NASA, the White House press corps), the receivers of messages (the U.S. public), and the context in which signification takes place (the object of study in economics, sociology, political science, philosophy). Semiotics leads us to questions about these things, but it cannot help us to answer them.

SAMPLE CRITICAL ANALYSIS: ONE MINUTE OF THE COSBY SHOW

What follows is a semiotic analysis of the opening credit sequence for *The Cosby Show* during the 1985–86 television season. The sequence lasts only a minute, yet it has been difficult to limit the analysis to a few pages. This is typical of semiotics, which often produces a discourse about an individual text that is many times longer than the original; because of this semiotics has gained a reputation for being rather tedious. The *Cosby* credit sequence has been chosen because it is something that may have been seen repeatedly, not just by the semiotician, who must go over the text a huge number of times in order to analyze it, but also by the average viewer of *The Cosby Show*. (See chart 1-1 for a shot breakdown and description of the sequence.)

CHART 1-1
The Cosby Show: Opening Credits Sequence, 1985–86 Season

Image
  denotation

SIGNIFIER	SIGNIFIED
Computer graphic resembling theater marquee with neon lights: "Let's All Be There" fade to black fade up	NBC logo

Close-up; one shot; medium gray background; key light from left.

Bill Cosby looks straight into camera, grinning, then raises eyes as he rocks his head from side to side in time to music as he exits frame left. (*Music begins*)

Graphic
White block letters on medium gray background.
Bill Cosby   The Cosby Show
in   in stereo
              where available
(*Main melodic line of music played by brass instruments begins*)
Close-up; one shot; medium gray backdrop; no camera movement.

Cosby enters frame right leading Phylicia Rashād (Claire Huxtable) by the hand. Cosby kisses her hand, exits frame left as Rashād enters frame right. She looks after him off screen, then looks at camera, smiling, as she wags her finger after Cosby.

Long shot; two shot; medium gray limbo backdrop. Strong key light on left casts shadows towards right edge of frame.

Rashād dances in front of Cosby, shaking her hips as she looks at him; then raises her arms and looks ahead. Cosby dances bent over, taking small steps, "clowning."

Rashād wears yellow blouse, black skirt, red shoes, earrings, belt. Cosby wears gray slacks, sweater with blue, red, green, black stripes.

Graphic   over image
          starring
          P H Y L I C I A   R A S H Ā D
                                    (*Music continues*)
Image
  denotation

S I G N I F I E R

S I G N I F I E D

Close-up; two shot.

Cosby and Sabrina LaBeauf (Sondra Huxtable) face camera, standing shoulder to shoulder; Cosby holds himself bolt upright. LeBeauf begins

CHART 1-1 (*Continued*)
The Cosby Show: Opening Credits Sequence, 1985–86 Season

---

	nodding her head in time to music, smiles. Cosby looks at her then back at the camera as he rolls his eyes, knits his brow.
Long shot; two shot.	LeBeauf dances in front of Cosby; Cosby dances bent over, takes tiny steps.

LeBeauf wears red shirt, black skirt, red boots.

Graphic   over image
              SABRINA LEBEAUF

Close-up; two shot.	Cosby and Lisa Bonet (Denise Huxtable) in profile; they look at each other, smiling, then at camera, then at each other as Bonet laughs.
Long shot; two shot.	Cosby and Bonet dance side by side; Bonet turns her back to him and "bumps" him off screen left.

Bonet wears wide red pants, gray jacket with epaulets; black shoes.

Graphic   over image
              LISA BONET

Close-up; two shot.	Cosby enters frame right leading Malcolm-Jamal Warner (Theo Huxtable) through frame; they both move their hands in front of their faces to the music
Long shot; two shot.	Cosby watches Warner dance vigorously, waving arms, jumping, touching floor with his hands; Cosby taps foot gingerly.

Warner wears black jeans, black and blue patterned sweater.

Graphic   over image
              MALCOLM-JAMAL WARNER
                      (*Music continues*)

Close-up; over-the-shoulder shot.	Tempestt Bledsoe (Vanessa Hux-

	table) tips her head from side to side in time with music, looks from the camera to Cosby as she stops smiling and moving. Cosby turns around to look over his shoulder at camera, glares, then turns again.
Long shot; two shot.	Bledsoe dances in front of Cosby. Cosby dances holding onto his pants as in a mock curtsy.

Bledsoe wears blue leather skirt, red shoes, gray, blue, red patterned sweater.

Graphic over image	
TEMPESTT BLEDSOE	
Close-up; one shot.	Keshia Knight Pulliam (Rudy Huxtable) dances, jumping so fast her image is blurred.
Long shot; two shot.	Cosby watches Pulliam dance, she jumps up and down rapidly, he takes tiny steps.

Pulliam wears red jumpsuit, red shoes, blue scarf tied at hips.

Graphic over image	
KESHIA KNIGHT PULLIAM	
Close-up; one shot.	Cosby nods head to music, smiles, eyes raised.
Long shot; one shot.	Cosby dancing, bent over, moving slowly, "clowning."

Graphics over image	
Associate Producer Caryn Sneider	
Creative Consultant Elliot Shoenman	
Producer Carmen Finestra	
Co-Executive Producer John Markus	
Close-up; one shot.	Cosby as he turns head quickly towards camera smiling.
Fade to black.	*(Music ends in time with his turn to the camera.)*

The Paradigmatic

Semiotics argues that the meaning of every sign derives in part from its relationship to others with which it is associated in the same sign system. Some of the meaning of this sequence, then, derives from its *difference* from the credit sequences of other TV shows. In the *Cosby* credits no speech is used (spoken or sung); no graphics appear other than titles; no special effects of any kind, such as split screens, freeze frames, or mattes are used; the characters look directly into the camera lens; none of the shots consists of takes from episodes of the program; none of the shots depicts the program's setting. These characteristics form the basis of a connotative meaning, which stems from the paradigmatic association with other program's credit sequences: *The Cosby Show* is different from the rest of television. As its producers (who, surely greet the weekly ratings in 1986 with glee) probably recognize, the connotation, "different and *better* than the rest of television," seems to have firmly taken hold with the U.S. public.

Four kinds of shots are used: close-ups and long shots of Cosby; close-ups and long shots of Cosby with one other family member. The lighting, camera angle, focal length, and background remain the same throughout the sequence. Such simplicity is again atypical of credit sequences, which usually consist of a variety of shots edited together in montage style. The gray, "limbo lighting" backdrop is unusual for television, but appears frequently in fashion photography, where it is thought to "set off" the models and their clothes (and offers no competition for the consumer's visual attention). Fashion is associated with having money to spend and with an aesthetic sensibility. Thus, the use of the gray backdrop in the *Cosby* credit sequence produces the connotations "well off" and "aesthetically inclined" and links these to the Huxtable family.

The Syntagmatic

The credit sequence has a simple structure: it begins with Cosby alone, introduces the family members one by one in relation to him, and ends with Cosby alone. A variation in shot composition occurs with the introduction of each family member, just as each one's facial expression, gestures, dance, and dress are "unique." This syntagmatic chain creates the sense of "a family of individuals," one of the show's principal themes. While Mrs. Huxtable and the children dance in a manner expressive of their personalities, that is, in character, Cosby/Dr. Huxtable clowns throughout the sequence—his dance is parody. The shots provide a vignette of Cosby/Huxtable's relationship with each family member: sexy-naughty, as he kisses Claire's hand and she wags her finger after him; mimicking adolescent seriousness with Sondra and Vanessa; enjoying Theo's exuberance and "cool." The sequence reinforces the sense of a family of individuals whom the father treats individually. Good-natured, thoughtful fatherhood and fun, playful family life are represented with great economy in the credit sequence.

Cosby's appearance throughout the sequence lends him a privileged status, which is reinforced by the use of Cosby's name throughout the credits and our knowledge of him as a "star." Mrs. Huxtable appears in the same syntagmatic relation to Cosby as do the family's children. The children also "belong" to Cosby in terms of the color code: he wears a sweater (the uniform of domesticity worn by all situation comedy fathers, but rendered particularly fashionable and expensive looking on Cosby, who never seems to wear the same clothes in two episodes) whose colors are precisely those worn, in different combinations, by the five children. The syntagmatic structure of the opening credits might be described as a theme and variations, where Cosby is the theme and each child—and his wife—appear as variations.

Generic Codes

Television situation comedy depends on a set of characters who live and work together, who are strongly differentiated in terms of "personality." Plots derive from absurd predicaments and good-natured power struggles that result from sharing quarters. Slapstick, pantomime, and verbal wit are its primary comic devices. As in other family situation comedies—*I Love Lucy, Leave it to Beaver, The Dick Van Dyke Show, My Three Sons, The Brady Bunch, All in the Family*—much of the humor on *The Cosby Show* derives from challenges to paternal authority or the "battle of the sexes." The opening credits sequence identifies Cosby as the central paternal figure, while it establishes him as the jokester, rather than the butt of jokes.

Perhaps most striking about the sequence is the fact that it is a dance number and therefore accumulates for the situation comedy some of the connotations associated with another genre, the musical comedy. Individual episodes of *The Cosby Show* frequently include musical numbers—one of the ways the show references black culture—that employ generic codes associated with a utopian sensibility.[18] Musical numbers allow for a kind of perfect, transparent expression of emotion, where characters communicate feelings directly, better than they are able to do with words alone. *The Cosby Show* is itself a utopian representation of the family: money is no object; love and harmony is the rule; play abounds as a means of solving discipline problems; marriage is sexy; gender equality is the stated goal; parents and children enjoy stimulating, satisfying situations at work and in school; childcare and housework are either invisible or enjoyable. Like a musical, *The Cosby Show* presents the world not as it really is, but as it should be.

Beyond the Text

Semiotics frequently speaks of a text as though it were understood in precisely the same way by everyone. At worst, it operates as though all meanings were translatable and predictable through the work of a gifted semiotician. Such an approach is nowhere more deficient than in television criticism, where the text is ephemeral but the audience remains a central preoccupation. We cannot, for example, speak of *The Cosby Show* without asking how a black program gained record-breaking popularity with an audience that, as measured by the Nielsen ratings, is overwhelmingly white. Because our society is characterized by racial segregation, white racism, and staggering percentages of black unemployment and poverty, *The Cosby Show* must also be examined in terms of the extratextual (television representations of whites and blacks; publicity about Bill Cosby and the show; racist discourses in other mediums) and in terms of the differences among its viewers that are structurally reinforced (gender, class, and race).

Semiotics can most usefully be seen as a descriptive method. Other approaches—feminist, psychoanalytic, ideological—can employ semiotics as a tool to ask larger questions of the television text. A feminist analysis might ask why the adult woman occupies a similar structural position to the children in her relationship to the adult man in this opening credit sequence, or why Bill Cosby has become so much more famous than any black actresses or comediennes on television. A psychoanalytic reading might question the privileged status of the father as producer and controller of the family discourse on *The Cosby Show*. An ideological analysis might ask what makes the Huxtable family— upper-middle-class, urban, predominantly female, and young—so popular with and acceptable to a white audience, or how the utopianism of *The Cosby Show* and its celebration of black achievement in the arts relates to the political struggle for racial equality.

NOTES

1. Umberto Eco, *A Theory of Semiotics* (Bloomington: Indiana University Press, 1976), p. 16.

2. Ibid.

3. Ibid., p. 17 (Eco's italics).

4. See the description of television coverage at the White House in Thomas Whiteside, "Standups," *The New Yorker*, 2 December 1985, pp. 81–113.

5. Eco, *Theory*, p. 7.

6. Herbert Zettl, *Television Production Handbook* (Belmont, Calif.: Wadsworth Publishing Company, 1984), p. 596.

7. Christian Metz, "On the Notion of Cinematographic Language," in *Movies and Methods*, ed. Bill Nichols (Berkeley: University of California Press, 1976), p. 586.

8. John Ellis, *Visible Fictions: Cinema, Television, Video* (London: Routledge and Kegan Paul, 1982), p. 130.

9. Roland Barthes, "Rhetoric of the Image," in *Image/Music/Text*, trans. Stephen Heath (New York: Hill and Wang, 1977), p. 38.

10. Ibid., p. 40.

11. Rick Altman, "Television/Sound" (paper delivered at the Twenty-fourth Annual Meeting of the Society for Cinema Studies, Madison, Wis., 24 March 1984), p. 13.

12. Umberto Eco, "Innovation and Repetition: Between Modern and Post-Modern Aesthetics," *Daedalus* 114 (Fall 1985): 162.

13. Eco, *Theory*, p. 214.

14. Christian Metz, *Film Language: A Semiotics of the Cinema*, trans. Michael Taylor (New York: Oxford University Press, 1974), p. 106.

15. Terry Eagleton, *Literary Theory: An Introduction* (Minneapolis: University of Minnesota Press, 1983), p. 109.

16. See Roland Barthes, "Myth Today," in *The Barthes Reader*, ed. Susan Sontag (New York: Hill and Wang, 1982), pp. 93–149.

17. Zettl, *Television Production Handbook*, p. 332.

18. See Richard Dyer, "Entertainment and Utopia," in *Genre: The Musical*, ed. Rick Altman (London: Routledge and Kegan Paul, 1981), pp. 175–89.

FOR FURTHER READING

A good place to start in the literature on semiotics is Terence Hawkes, *Structuralism and Semiotics* (Berkeley: University of California Press, 1977); or, for the more ambitious reader, Kaja Silverman, *The Subject of Semiotics* (New York: Oxford University Press, 1983). After that one might tackle some of the primary texts in the field: Ferdinand de Saussure, *Course in General Linguistics*, trans. Wade Baskin (New York: McGraw-Hill, 1966); Umberto Eco, *A Theory of Semiotics* (Bloomington: Indiana University Press, 1976); and Roland Barthes, *Elements of Semiology*, trans. Annette Lavers and Colin Smith (New York: Hill and Wang, 1968). Two other books by Roland Barthes that are particularly relevant are the essays collected in *Image/Music/Text*, trans. Stephen Heath (New York: Hill and Wang, 1977); and *Mythologies*, trans. Annette Lavers (New York: Hill and Wang, 1972).

Film has been scrutinized more carefully by semioticians than has television to date; some central works that may prove useful are: Christian Metz, *Film Language: A Semiotics of the Cinema*, trans. Michael Taylor (New York: Oxford University Press, 1974); Metz, *Language and Cinema*, trans. Donna Umiker-Sebeok (The Hague: Mouton, 1974); Jurij Lotman, *Semiotics of Cinema*, Michigan Slavic Contributions no. 5 (Ann Arbor: University of Michigan Press, 1976). For one perspective on the relationship of film semiotics to feminist criticism, see Teresa de Lauretis, *Alice Doesn't: Feminism, Semiotics, Cinema* (Bloomington: Indiana University Press, 1982); for a discussion of semiotic issues in ideological analysis, see Bill Nichols, *Ideology and the Image* (Bloomington: Indiana University Press, 1981).

A good introduction to semiotics in television criticism may be found in John Fiske and John Hartley, *Reading Television* (London: Methuen, 1978); the differences between film and broadcast television are provocatively laid out in John Ellis, *Visible Fictions: Cinema, Television, Video* (London: Routledge and Kegan Paul, 1982). Semiotics has been applied to many different kinds of television programs; the journals *Screen, Journal of Film and Video*, and *Jump Cut* are good sources of television criticism. Several important critical articles that use semiotics appear in E. Ann Kaplan, ed., *Regarding Television—Critical Approaches: An Anthology*, American Film Institute Monograph Series, vol. 2 (Frederick, Md.: University Publications of America, 1983). Phillip Drummond and Richard Paterson, eds., *Television in Transition* (London: British Film Institute, 1985) includes essays on television, from an international perspective, that evaluate the usefulness of semiotics for criticism and for understanding how audiences decode television.

Some other essays that raise interesting questions about the semiotic analysis of television include: Margaret Morse, "Talk, Talk, Talk—the Space of Discourse in Television," *Screen* 26, no. 2 (March/April 1985): 2–15; Janet Woollacott, "Messages and Meanings," in *Culture, Society and the Media*, ed. Michael Gurevitch et al. (London: Methuen, 1982), pp. 91–111; Nick Browne, "The Political Economy of the Television (Super)Text," *Quarterly Review of Film Studies* 9 (Summer 1984): 174–82; and Dennis Giles, "Television Reception," *Journal of Film and Video* 37 (Summer 1985): 12–25.

NARRATIVE THEORY AND TELEVISION
SARAH RUTH KOZLOFF

Whereas our ancestors used to listen to tall tale spinners or wise medicine men, read penny dreadfuls or the latest serialized novel, tune in to radio dramas or rush to the local bijou each Saturday, now we primarily satisfy our ever-constant yen for stories by gathering around the flickering box in the living room. Television is the principal storyteller in contemporary American society.

But what kind of storyteller is it? In what ways are stories presented on television similar to those transmitted through other media? What factors can we isolate as specific to television narratives? How can approaching television as a narrative art deepen our understanding of individual shows or of the medium as a whole? How can looking at television help us with our research on narrative itself?

The same decades that have brought the gradual invention, birth, and increasing maturity of broadcast television have also (totally coincidentally?) played host to the development of a new critical field, "narratology" or, more simply, "narrative theory." This field was founded in the late 1920s by the work of Vladimir Propp and the Russian formalist critics; it has since been fed by the studies of a diverse, international group of linguists, anthropologists, folklorists, literary critics, semiologists, and film theorists. Through their different perspectives on a variety of texts, a general outline of narrative structure and process has emerged and won a tentative consensus. Two books that summarize and synthesize the fundamentals of narrative theory—Seymour Chatman's *Story and Discourse: Narrative Structure in Fiction and Film* and Shlomith Rimmon-Kenan's *Narrative Fiction: Contemporary Poetics*—provide more detailed explanations of key concepts than is possible here. The present task is to use the fruits of this theory to open a discussion on the nature of television narratives.

Perhaps it is best to acknowledge that narrative theory has its limitations. Because this field is concerned with general mappings of narrative structure, it is inescapably "formalist" and largely unconcerned with questions about "content" and thus with political or ideological judgments. Similarly, because narrative theory concentrates on the text itself, it leaves to other critical methods questions about where the story comes from (for instance, the history, organization, and regulation of the broadcast industry, the intentions of the networks or of individual professionals) and the myriad effects (psychological or sociological) that the narrative has upon its audience. Later chapters will demonstrate critical approaches that fill in these large voids.

Yet, at the same time, we must not underestimate the importance of this critical vantage point, because American television is as saturated in narrative as a sponge in a swimming pool. Most prime-time television forms—the sitcom, the action series, the prime-time soaps, the made-for-TV movie, the feature film—are obviously narrative, as are such daytime offerings as cartoons and soap operas. Moreover, forms that are not ostensibly fictional entertainments, but rather have other goals—description, education, persuasion, exhortation, and so on—covertly tend to use narrative as a means to their ends. On the evening news, an unembellished recital of the latest economic figures is merely informative, but the *story* of the capture of a criminal uses narrative just as much as any cop show. A pain reliever commercial may rely upon comparison and argument, or a jazzy car commercial may be primarily abstract and poetic, but perhaps half of all advertisements find that the best way to sell their product is by offering a ministory exemplifying its beneficial effects. Rock videos range from straight dramatic performance, to weird expressionism, to those that enact the storyline of the ballad. Nature documentaries tend to follow the story of the life cycle of the animal or the geographic area through the passage of the seasons. The only television formats that consistently eschew narrative are those that are highly structured according to their own alternate rules: game shows, exercise shows, sports contests, news conferences, talk shows, musical performances. Yet even in such forms one can often spot narrative infiltration: talk show guests frequently answer their host's questions with stories about their past activities; football games are actually both played out by the teams and simultaneously narrated by sports announcers.

An episode of *60 Minutes*, "The Streets of San Diego," illustrates this habitual, covert "slide" into narrative. The thesis of the piece, which is openly stated by Harry Reasoner at the beginning, is that San Diego has the most compassionate policy in the country for dealing with the plight of the homeless. Accordingly, one might expect that the thrust of the piece would be descriptive ("*this* is what San Diego does for its homeless") and comparative ("*this* is how New York city handles its homeless"). But instead of such a dry expository approach, the episode actually proceeds through the following series of interlocking mininarratives:

Story 1: "History of 60 *Minute*'s Treatment of Homelessness"
   —flashback to 1982 episode that showed desperate plight of the homeless and insensitivity to their needs (scenes from Manhattan, Brooklyn, and Oregon)
   —in 1985 60 *Minutes* set out to find example of more compassionate treatment
   —found San Diego
Story 2: "The Education of Harry Reasoner"
   —as a midwesterner, his knowledge of San Diego was limited and clichéd
   —research on this story opened his eyes
Story 3: "Case History: Dennis Alexander"
   —Alexander arrived in S.D. from Chicago homeless, destitute, jobless
   —private charity got him work cleaning the streets
Story 4: "Flashback: How S.D. Community Leaders Developed Compassion for the Homeless"
   —group of seven community leaders recount how they went on the streets dressed as bums
   —they were not recognized by closest friends
   —learned rigors of the life firsthand
   —held news conference on return to civilization to dramatize problem
Story 5: "Flashback: Story of San Diego's Urban Redevelopment"
   —downtown formerly offered low-cost housing
   —urban renewal displaced the poor
   —city faced choice: take care of the poor or try to drive them out
Story 6: "Personal History of Father Joe Carol"
   —Catholic priest from the Bronx
   —gift for public relations noticed by his bishop
   —put in charge of fund raising for new St. Vincent de Paul shelter for the homeless that will be lush, not dehumanizing
   —had to learn to schmooze with rich donors at cocktail parties

Harry Reasoner provides a bit of expository glue to link these stories together, but as one can see, this episode consistently turns to storytelling.

The viewer watching "The Streets of San Diego" is likely to have been highly entertained and slightly edified; one learns a bit about San Diego, about the connection between urban renewal and homelessness, and about how the actions of a few responsible community leaders can raise the consciousness of a community. But what is interesting about this text is what was left out: the viewer is left with very few specifics on who helped Dennis Alexander and how, on who is in charge of policy for the homeless, on how government and religious groups are cooperating, or on what is being done for the needy while the shelter is under construction. Nor is one given any comparisons as to how other cities are currently coping with the problem.

Thus, narrative seems to have run away with the 60 *Minutes* team, and to have swept the viewer along too. My point is not that narrative is an evil distraction from careful reasoning (though obviously it can be), but rather that on television, just as in other media, it is both omnipresent and uniquely powerful. Yet we are so used to it and it seems so natural that we generally allow it to pass unmarked. The solution, of course, is to step out of its magnetic pull and take it apart piece by piece.

To this end, we learn from narrative theory that every narrative can theoretically be split into two parts: the "story," that is, "what happens to whom," and the "discourse," that is, "how the story is told." To recognize television's specificity, I believe we need to add a third layer, "schedule," that is, "how are the story and discourse affected by the narrative's placement within the larger discourse of the station's schedule." Let us begin with the innermost layer.

STORY

Rimmon-Kenan defines a story as a "series of events arranged in chronological order." She correspondingly defines an "event" as a "change from one state of affairs to another."[1] Tzvetan Todorov uses different terms, but he is talking about the same phenomenon when he defines a minimal narrative as a move from equilibrium through disequilibrium to a new equilibrium.[2] For example, in "The Streets of San Diego," Harry Reasoner was ignorant about the city ("equilibrium" or "state of affairs"), then he learned more about it through doing the episode ("disequilibrium" or "event"), and now he is enlightened ("new equilibrium" or "new state of affairs"). Rimmon-Kenan's and Todorov's definitions do not quite make explicit the fact that events cannot occur in a vacuum—they must be enacted by a given set of characters (Harry Reasoner) in certain settings (the midwest, San Diego). Seymour Chatman groups characters and settings under the label "existents." Together, "events" and "existents" are the basic components out of which stories are made.

Although events and existents are inseparable, we may begin by concentrating on the former. Out in the real world, things can happen totally at random, but in stories they are linked by temporal succession (X occurred, then Y occurred) and/or causality (because Y occurred, Z occurred). One of the things that television, like all other narrative forms, takes advantage of is the viewer's almost unquenchable habit of inferring causality from succession. For example, a simple Nyquil commercial first shows a man and a woman together in a double bed, both snuffling and sneezing. (We understand them to be husband and wife, afflicted with horrible colds.) Without dialogue, the woman takes some Nyquil from her bed table and offers it to the man; he declines, taking another medication. A title reads LATER; then we see the woman fast asleep, while the husband is still miserably awake. Note that the commercial links these two scenes merely by an indicator of temporal succession, "later," but the advertisers know full well that the viewer will make a causal connection: the wife is sleeping peacefully *because* she took Nyquil.

Not all story events are of equal importance. As Roland Barthes was the first to point out,[3] one can determine a hierarchy between the events that actively contribute to the story's progression and/or open up options (Chatman labels these "kernels") and those events that are more routine or minor ("satellites"). In the Nyquil commercial, the important, kernel event is the decision to take the medication or not to take it; "sitting up in bed," "reaching for the bottle," and "unscrewing the cap" may be events, but they are minor satellites.

Story events are not arranged at random. For millennia, one of the tasks of critics has been the discovery and detailing of stories' underlying structures. It was Aristotle who first pointed out the seemingly banal but actually vital fact that the plots of tragedies have a beginning, a middle, and an end, and that they neither begin nor end haphazardly.[4] Over a century ago, German playwright and novelist Gustav

Freytag elaborated on this insight by describing the typical "dramatic triangle": well-made plays begin with an expository sequence setting out the state of affairs, rise through various twists and turns of complicating actions to a climax, and then fall off in intensity to a coda that delineates the resolution of the crises and the new state of affairs.[5] The majority of television stories follow exactly these patterns; however, certain shows—serials and soap operas—break ranks by resisting resolutions and refusing to end. Serialization will be discussed in greater detail when we look at scheduling.

Because any series of events can be strung together to make a story, theoretically the number of possible stories to be told is infinite, and each story could differ radically from all others. In practice, however, narrative theorists have discovered that many, if not most, stories fall into predictable, discernible patterns.

In a pathbreaking study, *Morphology of the Folktale*, first published in 1928, Vladimir Propp studied a group of Russian fairy tales. He invites the reader to compare the following events:

1. A tsar gives an eagle to a hero. The eagle carries the hero away to another kingdom.
2. An old man gives Súčenko a horse. The horse carries Súčenko away to another kingdom.
3. A sorcerer gives Iván a little boat. The boat takes Iván to another kingdom.
4. A princess gives Iván a ring. Young men appearing from out of the ring carry Iván away into another kingdom.[6]

Obviously, something uncannily similar is going on here.

Propp concludes that although different tales may feature different characters, these characters fall into one of seven types of dramatic personae: hero, villain, donor, dispatcher, false hero, helper, and princess and her father. Moreover, despite surface variability, the actions of these personae serve identical purposes in terms of their "function" in moving the story along. Propp thus was able to formulate the following laws:

1. Functions of characters serve as stable, constant elements in a tale, independent of how and by whom they are fulfilled. They constitute the fundamental components of a tale.
2. The number of functions known to the fairy tale is limited.

3. The sequence of functions is always identical; and
4. All fairy tales are of one type in regard to their structure.[7]

Propp found himself able to compile a list of thirty-one functions occurring in his tales. These tales trace a hero's quest and/or contest
with a villain; thus typical functions include such activities as "#6:
The villain attempts to deceive his victim in order to take possession of
him or of his belongings," and "#12: The hero is tested, interrogated,
attacked, etc., which prepares the way for his receiving either a magical agent or helper."[8] Propp's list of functions specifies all the different
categories of events found in these tales and the sequence in which
they transpire.

Certain cartoons—for example, She-ra, Princess of Power and He-
Man and Masters of the Universe—which imitate traditional fairy
tales and rely on such motifs as sorcery and disguise, exhibit blatant
parallels to Propp's model. But let us consider also the following, very
deliberately crafted twentieth-century stories:

1. Housewife X's sink is clogged. Josephine the plumber suggests
   Liquid Plumr. The drain cleaner cuts through the clog and the
   housewife's problem is solved.
2. Customer Y has dry, chapped hands from washing dishes.
   Madge the manicurist suggests Palmolive dishwashing
   detergent. Customer Y returns to beauty parlor with restored
   hands.
3. Housewife Z makes bad coffee and husband complains. Mrs.
   Olson recommends Folger's coffee. Housewife Z tries Folger's
   and wins husband's praise and affection.

In each of the above stories, the heroine has a lack or misfortune
(Propp's function #8a), which is noticed (#9). She comes into contact
with a donor (#13), who suggests the use of the magical agent (#14).
The initial misfortune or lack is liquidated (#19). Often the heroine is
then praised and thanked by family members (figuratively #31, "The
hero is married and ascends the throne").

Longer, more complicated television narratives don't exactly correspond to Propp's model (although film theorist Peter Wollen is quite
successful in applying the model to Alfred Hitchcock's North by North-

west).[9] Nevertheless, television characters frequently fall effortlessly into Propp's categories of dramatic personae, as the following table illustrates.

New villains appear every week so that the hero has a fresh challenge; by the same token, princesses generally vary from week to week so that each show can spark romantic interest without the hero getting tied down. On the other hand, donors and false heroes belong more to Faerie than to American action series—examples are hard to come by. (Perhaps Oscar Goldman and Dr. Rudy Wells qualify as donors; they "donate" Steve Austin's and Jamie Somer's magical abilities on the *Six Million Dollar Man* and *The Bionic Woman*. As for the false heroes, Remington Steele may qualify—he is always stealing the credit for Laura Holt's detecting—but because, instead of doubling as villain, Remington actually doubles as the prince, the stories always carefully protect him from exposure and deny Laura her deserved recognition.)

One may draw two conclusions from the applicability of Propp's work to television. The first is that American television is remarkably like Russian fairy tales—that is, that certain motifs, situations, and stock characters may have a nearly universal psychological/mythological/sociological appeal and thus appear again and again in popular cultural forms. (An intriguing prospect, but off our current subject.) The second conclusion is the one drawn by subsequent narrative theorists, namely, that stories are governed by a set of unwritten rules acquired by all storytellers and receivers, much the way we all acquire the basic rules of grammar. This conclusion explains both stories' variability and consistency; a sentence can be composed from an almost infinite choice of subjects, verbs, and objects, but to be comprehended, these choices must be arranged according to certain shared rules. One major strand of narrative theory since Propp has accordingly concentrated on further specifying these rules and ironing out Propp's flaws and inconsistencies; unfortunately, the schema drawn up by Tzvetan Todorov, A. J. Greimas, Gerald Prince, and others also have drawbacks.[10]

It seems to be true that popular cultural forms are more rigidly patterned and formulaic than works of "high art." Thus television shows that are not particularly "Proppian" still display discernible armatures. One can practically guarantee that each week on *Star Trek* the USS *Enterprise* will encounter some alien life form, members of the crew

CHART 2-1
Television Characters and Propp's Dramatic Personae

SHOW	HERO	HELPER	DISPATCHER	VILLAIN	PRINCESS
Batman	Batman	Robin	Police Comm. Gordon	The Penguin The Riddler etc.	
Man From U.N.C.L.E.	Napoleon Solo	Illya Kuryakin	Alexander Waverly	T.H.R.U.S.H.	guest star
Starsky & Hutch	Starsky & Hutch	Huggy Bear	Capt. Dobey	guest star	guest star
Miami Vice	Crockett Tubbs		Lt. Ortega	guest star	guest star
Spenser: For Hire	Spenser	Hawke	Lt. Quirk	guest star	Susan Silverman

will be separated from the ship (which will itself be placed in jeopardy), one crew member will have a romantic interest, all will be resolved through the crew's resourcefulness or high-mindedness. Each week on *Bewitched,* Samantha's magical powers will get her into some tangle potentially leading to exposure of her witchcraft and/or great embarrassment to husband Darren; said problem will be remedied in the nick of time. On *Perry Mason,* Perry will take a case that looks hopeless while the prosecutor gloats; Della Street and Paul Drake will uncover crucial bits of evidence; Mason will break down witnesses on the stand, exonerate his client and uncover the real killer. More generally, one can rest assured that harmony will be restored at the end of each sitcom, children will be taught a moral lesson at the end of each domestic comedy, crime and detective shows will include a chase, and dastardly villains will inevitably be vanquished by their own dastardly inventions.

Such predictability had led scholars to remark on television's deficiencies in terms of one of the major engines driving narrative—suspense. As Roland Barthes argues in *S/Z,*[11] each significant event opens up a number of possibilities; the reader or viewer is constantly in a state of suspense and anticipation, wondering "what next? what next?" One does find a few genuinely suspenseful and unpredictable stories on television, primarily on newscasts and news documentaries. For instance, in addition to their moral and political significance the Watergate and Iranian hostage crises were compelling as stories; each evening broadcast brought complicated, unpredictable twists and turns, and it was by no means certain that the good guys were going to win out. But because so much of fiction television is so formulaic, and because we know that (except in the case of special broadcasts) the hero or heroine will be back next week, we are hardly ever in the real suspense one might feel with a film or novel as to whether the hero and his love interest will triumph, or even survive.

One means by which television narratives compensate for their lack of suspense is by proliferating storylines. Often a show will use the same protagonist for separate storylines, as when action shows involve their heroes in both a case and a romance. Other shows will use different family members as the leading players in two or three separate storylines; soap operas will keep as many as five or six storylines, involving different characters, hopping simultaneously. Each given storyline may be totally formulaic, but the ways in which it combines with, parallels, contrasts, or comments upon another storyline can be totally unique.

An episode of *Magnum, P.I.,* "Old Acquaintance," illustrates the ways in which television shows frequently multiply and connect storylines. The show begins by showing the theft of a dolphin from an outdoor aquarium. Next we see Magnum at his home with T.C. and Rick. Magnum is preparing to meet an old acquaintance. Goldie Morris had been a friend of his in high school and had tutored him to help him pass English. T.C. and Rick tease Magnum about Goldie's unattractiveness in her high school photo. As Magnum is about to leave Higgins enters, asking Magnum to drop off his credentials at the yacht of a visiting president of a mythical African nation. Higgins is planning to attend a meeting of the International Human Rights Advisory Council on this yacht as an unofficial representative of the British government, which wants to keep its distance from President Kolé and his record of atrocities towards his people.

So in the first few minutes, three storylines are set in motion:

1. The theft of the dolphin. Who took it and why? Will it be recovered?
2. Magnum and Goldie's relationship. When it turns out that she's become quite pretty, will their friendship change to romance? and
3. Higgins's meeting with President Kolé. Should he go? What is to be done about Kolé's crimes?

It turns out, however, that these three storylines are intimately interconnected: Goldie was involved in the theft of the dolphin (out of overzealous concern for animal rights), but her politically extreme cohorts have actually kidnapped the dolphin to use it to carry a bomb to blow up Kolé and all aboard his yacht. Magnum recovers the dolphin, saves everyone's lives (assassination is not the answer to dealing with dictators) and, affectionately but paternally, extricates Goldie from her difficulties (thereby paying her back for her past kindnesses and reasserting his superiority). As in Victorian novels, coincidences abound and storylines coalesce into order and harmony.

Proliferating storylines diffuse the viewer's interest in any one line of action and spread that interest over a larger field. In general, I would extend Robert Allen's insight about soap operas to cover the lion's share of narrative television:[12] television stories generally displace audience interest from the syntagmatic axis to the paradigmatic, that is, from the flow of events, per se, to the revelation and development of "existents," or setting and characters.

One of the reasons why commercials are so glossy and slick is that, because the sponsor plans to air each commercial on numerous occasions, the high production value is necessary to recapture the viewer's attention when all suspense has been dissipated. Like commercials, certain high-budget television shows (*Love Boat* cruises to exotic locales, miniseries such as *Lace*) seek to make up for their vapid stories by careful camerawork of unusual scenery, gorgeous costuming and set direction.

But the average prime-time series has a relatively undistinguished setting; opening montage sequences may situate the show in a particular locale, but once the action begins, the living room, bedroom, office, restaurant, hospital studio sets are not particularly evocative or individualized. Only shows that go out of their way to include exteriors of local landmarks, such as *The Streets of San Francisco* or *Hawaii Five-O*, truly make much use of their settings.

In fact, as others have noted, it is characters and their interrelationships that dominate television stories. The way the medium presents characters contrasts markedly with the situation in literature; despite the apparent individuality and vibrancy of an Emma Bovary or Huckleberry Finn, literary and narrative theorists argue over whether literary characters can truly be said to exist. Extremists claim that it is nonsense to think of them as people; they are merely phantasms, nothing but concatenations of the actions they perform or the traits ascribed to them; ultimately Pip dissolves into nothing but words on a printed page. However, television narratives, like films, indisputably offer viewers *people*. We see Heathcliff Huxtable every time we watch *The Cosby Show*—he is a living, breathing person. Undeniably the extent to which Cliff exists apart from Bill Cosby may be open to debate (note that this show, like so many others, is named after the actor, not the character). But whether one thinks of the man on the screen as Bill Cosby or as Cliff Huxtable, no one can deny that it is his charm and personality, not the episodes' events, that draw viewers to the show.

Predictable as their events may be, television stories offer us a wide gallery of vibrant characters. The fact that these characters, as discussed above, can generally be slotted into certain common categories does not diminish their magnetism. Cliff Huxtable is the "Father figure" in a domestic comedy, and as such he fulfills certain set functions (dispenser of wisdom, disciplinarian, breadwinner, devoted husband), but he fulfills these functions quite differently than Mr. Cleaver does in *Leave It to Beaver.*

Moreover, as David Marc argues, each episode of a series contributes to the "broader cosmology of the series."[13] Television series nearly always create in their initial premise a tension or enigma that centers on character development or relationships. Will Mary Richards be able to make it on her own? Will Alex Keaton renounce greed and ambition and embrace more human values? Will Edith Bunker ever revolt against Archie? Numerous shows (such as *Gunsmoke, The Avengers, Cheers, Moonlighting*) thrive by exploiting the tension of covert or undeclared heterosexual passion—will Matt Dillon/John Steed/Sam Malone/David Addison and Kitty/Emma Peel/Diane Chambers/Madelaine Hays ever marry?

To take an example, the central question of *Magnum, P.I.,* as I see it, is "when will Magnum grow up?" No one really watches *Magnum* for its plot: one watches to see the emergence of the maturity, integrity, and competence Magnum hides underneath his boyish, careless manner; to see if he will win Father Higgins's respect; to see if he will renounce the company of his (even more adolescent) buddies for the ultimate proof of adulthood, marriage. "Old Acquaintance" engages all of these strands: Goldie is offered as a potential spouse, yet she is revealed as unsuitable; Magnum proves his abilities once again (even saving Higgins's life), yet gets to stay unattached. Each episode of a series toys with the series' enigma—embellishing it, replaying it, bringing it to the brink of resolution and then drawing back. (The closing state of affairs is therefore likely to be almost identical to the initial state of affairs.) Because we like the characters so much, and because the enigmas always center on things as psychologically potent as love, maturity, independence, etcetera, the surface plots of television shows, in a sense, become inconsequential.

Television stories may be formulaic, but the ways in which they are told can vary considerably. Thus, let us move on to look at narrative discourse.

DISCOURSE

Participants

Perhaps the key factor in approaching narrative discourse is to comprehend that the story is indeed *told*. On your way to the store you may witness a chronological series of events enacted by various personages in a given setting—say a purse-snatching and the apprehension of the thief—but what you have witnessed is not a narrative, it only becomes a narrative when you relate what you have seen to your friends. Narration is a communicative act; to have a narrative, one must have not only a tale, but also a teller and a listener.

A substantial portion of narrative theory has focused on studying the participants in this special exchange. As Robert Scholes and Robert Kellogg noted some years ago,[14] our model of narrative transmission comes from the days when one sat and listened to a physically present storyteller spin his fantasies. With the move to literary narratives, the situation became more complicated, because instead of actually listening to a storyteller, we have a printed text in which an author has deliberately inscribed an imitation storyteller, that is, the narrator. In fact, on a theoretical level, literary narratives always involve the following six participants:

TEXT

Real Author → | Implied Author → Narrator → Narratee → Implied Reader | → Real Reader

To (briefly) describe these six participants, let us pretend that the text under consideration is *Huckleberry Finn*. The "real author" is the flesh and blood writer, Samuel Clemens. The "implied author" is the imaginary conception of "Mark Twain" that a reader constructs from reading the text. (Because each reader formulates his or her own image of "Twain" from weighing subtle hints in the text, readers may not always agree on his characteristics; for instance, some argue that *Huckleberry Finn* leads them to believe that the person behind the work is terribly racist, others that he is a fierce critic of racism.) The "narrator" is Huck; he is explicitly set forth in the opening lines as the voice telling the tale: "You don't know about me without you have read a book by the name of *The Adventures of Tom Sawyer;* but that ain't no matter." The "narratee" is the unspecified person, that "you" above, to whom Huck is supposedly speaking. The "implied reader" is the imaginary reader for whom the implied author seems to be writing—someone, in this case, who is willing to criticize the cant and foibles of civilization. The "real reader" is the flesh and blood person reading the book in his or her armchair.

Because the above chart grew out of theorists' analyses of literary narratives, complications arise in applying it to film or television. Who, for instance, is the real author of the *Star Trek* series? The episodes were written and directed by a roster of different professionals. Because television producers typically have more continuing involvement and power than writers or directors, we might turn to Gene Roddenberry, the series' creator and executive producer, as the person preeminently responsible for it. But in general, as with cinema, television is a collaborative enterprise and it is difficult to assign to an individual the title and status of authorship.

The "implied author" of a television show, like that of a novel, is the authoring personality that the viewer constructs based on the choices manifest within the text. Many television shows are so conventionally scripted and shot that it is hard to get any definite sense of this figure, but in some cases one can form a general impression and make broad contrasts. Take, for example, three crime shows of three different producers: behind Stephen Bochco's *Hill Street Blues* one senses someone fatalistic, irreverent, and politically liberal; behind Quinn Martin's *The FBI* stands someone who believes in law and order and humorless professionalism; behind Levinson's and Links's *Murder, She Wrote* flits a lighthearted, antiaristocratic imp.

Because television is a collaborative media, questions of authorship are complicated, but because it is both aural and visual, questions of narration are even more so. Our model of a narrator is of a person speaking aloud, whereas television shows and films proceed instead through the unrolling of a series of moving images and recorded sounds. Yet we know that someone/something is presenting these images in just this way—someone/something has chosen just these camera setups and arranged them in just this fashion with just this lighting, these sound effects, and this musical score. As Christian Metz leads us to see, because it is narrative, someone must be narrating.[15] The real difficulty is in deciding what to call this intangible narrating presence; various terms have been proposed, including "grand image-maker," "implied narrator," "implied director." Perhaps the most common solution is just to use "camera" as a shorthand referring not only to the machine itself, but to all the markers of narration; thus, one will often hear phrases like, "the *camera* then cuts to another scene," or "the *camera* then zooms in."

Partly because the narrating presence behind a television show is impersonal and nebulous, time after time television naturalizes this strangeness by offering a substitute human face and/or voice. In the fifties the dramatic anthology series all seemed to have "hosts" who would appear before the story itself and act as introducers and emcees. The practice continued through the sixties in Rod Serling's and Alfred Hitchcock's introductions to *Twilight Zone* and *Alfred Hitchcock Presents,* and figures to this day in Alistair Cooke's hosting of *Masterpiece Theater.* These on-camera hosts lend their charms and credibility to the camera/narrator; they serve to personalize the impersonal.

In other cases the narrator is humanized merely through a disembodied voice, through voice-over narration. Commercials, of course, use voice-overs incessantly, as do documentaries, newscasts, and sports events. The voice works in tandem with the visual track; it is the "camera's" voice, telling us what we are seeing, or what to think about what we are seeing.

Fiction shows use voice-over narrators more frequently than one might at first realize. Some shows just utilize these narrators at the beginning, to set up the premise of the series, as in the ballad that introduces *The Beverly Hillbillies* ("Come and listen to a story 'bout a man named Jed/Poor mountaineer barely kept his family fed"). Others make the oral narration an integral, ongoing facet of the text. Basically, voice-over narrators break down into two types: anonymous narrators who are situated somewhere outside the world of the story they relate (for example, the narrators of *The Untouchables* and the "Fractured Fairy Tales" on *Rocky and Bullwinkle*) and character-narrators such as Joe Friday in *Dragnet* or John-boy in *The Waltons.* We will look at voice-over narrators more in the next section.

The concept of the "narratee" is particularly helpful for the study of television because, inasmuch as the shows are broadcast so widely to vast, impersonal audiences unknown to the shows' creators except as statistics, the authors have frequently resorted to using stand-ins. How many times has one heard, "Show X was filmed live before a studio audience?" Consciously or not, the networks invite these audiences to make the communicative act concrete—the story now is being told for real listeners (as opposed to video lenses), and the actors and director can get immediate feedback from the audience's reactions. Furthermore, the viewer isolated at home can now get the sense that he or she is experiencing the narrative communally, and his or her reactions are likely to be augmented by the example of the studio audience's reactions. Alternatively, shows will skip the trouble and expense of inviting an audience and instead substitute canned narratees in the guise of a laughtrack. In such cases, the track does not serve to enliven the performance but only to prompt the home viewer's responses.

Another type of television narratee is the "perfect listener." The visiting star on the *Tonight Show,* or *Dick Cavett,* or *Barbara Walters,* recounts the story of his or her career/drug/personal crises and recovery to Johnny Carson, Dick Cavett, or Barbara Walters. Similarly, news shows often find it preferable for the reporters in the field to address their stories not straight to the audience at home but rather to the anchorman/woman in the control booth. The talk show hosts and the anchorpersons fulfill identical functions—they listen eagerly, sympathetically, and ask intelligent questions. They are "perfect listeners," and their interest and attention is supposed to serve as a model for the viewer eavesdropping in on this conversation at home.

Less commonly, a television show will create a narratee who is intimately connected to the narrative's fictional world. *Star Trek* features abundant voice-over narration by Captain Kirk, or, if he is separated from the *Enterprise,* by whomever assumes command. This narration assumes the form of a "Captain's Log." Because the log is addressed to Star Fleet Command, this command serves as the ostensible narratee. For us, the viewers, to be hearing/watching the log puts us in an interesting position: either *we* are Star Fleet Command, or we are uniquely privileged to overhear Star Fleet's communications. Just by its choice of how to couch its narration, *Star Trek* subtly manages both to flatter and involve its viewers.

The "implied viewer" of television narratives is a fictional construct, the person who communes perfectly with the implied author. Thus, the implied viewer of *Hill Street Blues* believes in women serving on police forces and in interracial friendships; the implied viewer of *Murder, She Wrote* values small town, homey simplicity. Though it may seem self-evident on a practical level, it is worth noting in this theoretical context that Schlitz commercials are addressed to people who drink, not abstainers; Pampers commercials are addressed to parents, not bachelors; Cascade commercials to those who own diswashers (or plan to acquire one someday), not to those who don't. In short, each commercial creates an implied viewer who is interested in its message. Even if you don't own a dishwasher, when you watch a Cascade commercial, in order to meet the narrative on its own terms you must pretend that you do.

Finally, the "real viewer" is an unproblematic term; it refers to the flesh-and-blood viewers in their living rooms.

The distinctions between these six participants can help us in our study of television in various ways. First, we should be able to place our work precisely. Biographical studies of creative professionals deal with real authors, but a study that concentrates on the ways in which personality traits and moral viewpoints are manifested within a television show pertains to implied authorship. Statistical studies of audience responses analyze real viewers, but reader response criticism is more concerned with implied viewers. And there has, to my mind, been altogether too little work yet done on the figures at the heart of the exchange, the television narrator and narratee.

Furthermore, this model can help us understand a facet of television so often commented upon—the medium's propensity for "direct address." Direct address refers to the situation when someone on TV—a news anchor, a talk show host, a master of ceremonies, a reporter— faces the camera lens and appears to speak directly to the audience at home. In this situation what we seem to have is a precipitous collapse of the six narrative participants into merely two—the speaker and the viewer. When Dan Rather (who functions both as supervising editor and anchor of the *CBS Evening News*) faces the camera and relates the evening news, he simultaneously figures as real author, implied author, and on-screen narrator, while I, sitting at home, am simultaneously narratee, implied viewer, and real viewer. Although theoretically there is always a distinction between these roles—between, say, Rather as flesh-and-blood author, and Rather's screen persona as implied author—the distinctions in these cases are nearly indiscernible. Such a strong impression is given of a direct interpersonal exchange that when Rather says "Good night," I, for one, am likely to answer back to the screen, "Good night, Dan."

Whenever we get back to two participants, we are back to the original model of the prototypical narrative exchange—the storyteller and the listener. In *Regarding Television,* John Fiske and John Hartley refer to television's "bardic" function as the conveyor of a culture's values and self-image.[16] I would maintain, however, that it is also "bardic" in that, despite its technological sophistication, it frequently seeks to imitate the most traditional and simplest of storytelling situations.

Typology of Narrators

As I said above, very little work has yet been done on analyzing television narrators. Narrative theory can provide crucial help here in that it has already isolated a host of issues concerning the relationship of a narrator to his or her tale. These issues used to be classified under the term "point of view," but narrative theorists and literary critics currently recognize that the term is outdated and discredited—too many discrete variables have been lumped together and tangled up under this one heading. Because no wide-ranging consensus has yet been reached on how to divide and rename these variables, for our purposes here we will eschew terminological precision and concentrate instead on identifying six of the most important.

First, is the narrator a character in the story he or she tells, or is the narrator outside of the story-world (also called the "diegesis")?

I referred to this distinction briefly in an earlier section. Captain Kirk, Thomas Magnum, John-boy Walton are character narrators (in Gerard Genette's terms "homodiegetic"); Walter Winchell, who narrates *The Untouchables,* Edward Everett Horton, who narrates "Fractured Fairy Tales," and standard voiceless "camera-narrators" are exterior to the stories they relate ("heterodiegetic"). The distinction between these two kinds of narrators can be important because conventionally (not automatically) character narrators are considered less objective and less authoritative than heterodiegetic narrators. The former are personally involved in the stories they relate, the latter merely observing from some Olympian vantage point. (Television reporters, somehow, frequently manage to have it both ways. Harry Reasoner tells us about San Diego's efforts for the homeless from a position of objective authority and detachment; he is not involved in these efforts, he is a reporter. Yet at the same time he personalizes the story: he dwells on his own feelings and jokes with the people he interviews.)

Second, does the narrator tell the whole tale, or is his or her story embedded within a larger "framing" story?

Whenever a character within a series tells another character a story (as in a *Family Ties* episode in which family members sit around swapping embarrassing stories about Alex), these narrators are embedded within the overarching discourse of the camera-narrator. Embedded narrators have on-screen narratees, and the viewer at home is put in the position of an eavesdropper. They are usually less powerful, less objective than framing narrators. In contrast, framing narrators are responsible for the entire tale, and they generally address the viewer directly.

Third, what degree of distance in terms of space and time exists between the story events and the time and place of the narrator's narrating?

John-boy Walton narrates from the vantage point of a grown man; he is nostalgic and reflective. (John-boy is portrayed on-screen by Richard Thomas, but an older actor provides John-man's voice-over.) On the other hand, Jim Kirk narrates as his story unfolds. He is wrapped up in the action; his narration is much more anxious and immediate.

Fourth, what degree of distance in terms of detachment, irony, or self-consciousness does the narrator exhibit?

Most camera-narrators strive for neutrality and self-effacement, as if the viewers are supposed to overlook the fact that the story is coming to them through a mediator, and instead believe that they are looking in on reality. However, no one could be more ironic and self-conscious than Alfred Hitchcock in his introductions, or Edward Everett Horton in his narration of the "Fractured Fairy Tales." Such narrators poke fun at the characters in the story and at themselves as storytellers; yet they create a special kind of intimacy by inviting the viewer to share their joke.

Fifth, is the narrator reliable? If unreliable, does he or she withhold the truth through his or her own limitations or in order to mislead us?

The way to tell whether a narrator is unreliable or not is to look for discrepancies between what the narrator tells us and what we intuit the implied author believes. Voice-over narrators of commercials generally strive for perfect sincerity and every other facet of the text is designed to bolster their credibility. In "Old Acquaintance," by contrast there is a scene when Magnum goes to the hotel to meet Goldie. His voice-over states: "I had to admit I was a little nervous about seeing Goldie again after all these years. But one thing I wasn't worried about was whether I'd recognize her or not. There was a bond between us, a history, a camaraderie that went beyond the physical. It was a spiritual sort of thing." Meanwhile the shot shows Thomas craning around the lobby and overlooking a lovely redhead, Goldie, who is blatantly trying to attract his attention. This dichotomy shows us that Thomas has been spouting garbage; just like T.C. and Rick, he is hung up on "the physical," and his "spiritual bond" is not strong enough to overcome his memory of Goldie's high-school unattractiveness.

Finally, one might look at the narrator's degree of omniscience. Omniscience may involve one or more of the following traits: knowing the story's outcome, having the ability to penetrate into characters' hearts and minds, and/or having the ability to move at will in time and space. Most television narrators display a large degree of omniscience, but occasional significant variations can be registered. In some crime shows, such as *Hawaii Five-O*, the camera shows the viewer the guilty party at the onset; we side with the narrator in a position of knowledge and wait for McGarrett to catch the crook. In other cases, the camera-narrator, protagonist, and the viewer are all equally in the dark and we all wait patiently for the protagonist to solve the puzzle. In still other cases, the camera-narrator Knows All but resists Telling All; for instance, it shows the murder being committed but coyly keeps the murderer's face off-screen. In *Dallas*, the narrator knew who shot J.R.; it just wouldn't tell us until the following season.

In short, point-of-view studies entail looking carefully at the narrator's position vis à vis the tale and the consequences of this position to the discourse as a whole. Identical story events can seem radically different depending upon the narrator's point of view and on the degree of his or her power, remoteness, objectivity, or reliability. As Walter Benjamin once put it, "Traces of the storyteller cling to the story the way the handprints of the potter cling to the clay vessel."[17] Analyzing television narrators, then, involves putting a magnifying glass to these individualized handprints.

Time

Christian Metz has written, "There is the time of the thing told and the time of the telling. . . . One of the functions of narrative is to invent one time scheme in terms of another time scheme."[18] In the early days of television, when shows were broadcast live, it was possible to deny or overlook television's narrative component and align the medium with drama; after all, the time of the thing told and the time of the telling were congruent. Nowadays, however, when nearly all shows are videotaped, the existence of two time schemes is much more noticeable and undeniable. Even in the case of events such as live football games, the viewer does not—unlike the stadium ticketholders— simply witness the progression of the game. Instead she or he sees the events filtered through the control room, which switches from crowd shots, to the cheerleaders, to the coaches, to the action; which flashes back to pregame interviews; which forsakes real time for slow motion and freeze frames; and which repeats the same play over and over. The viewer is no longer simply watching the game, but rather a narration of the game in which various choices have been made concerning temporal order, duration, and frequency.

Story events, by definition, proceed chronologically. But when the teller tells the tale, she or he is not bound to follow chronological order; the events can be presented in any order the teller finds most effective. A narrator might choose to use a flashforward to entice the viewer with glimpses of the action or drama ahead; for instance, *Mannix* episodes begin with credits and theme music over a montage of shots of Mannix engaged in various exciting/dangerous activities; these shots are snippets of the particular dangers the hero will be facing on this evening's episode. Or, a narrator might employ a flashback to orient the viewer and bring him or her up to date; news stories often intercut file footage from previous events to educate the viewer, and serials often begin with a montage of scenes from earlier shows.

One particularly urgent issue that a television narrator faces in terms of ordering events is how best to deal with simultaneity. As mentioned above, television stories rather frequently encompass more than one storyline; in the story-world these events may be happening at the same moment, but a narrator can only narrate one thing at a time. Before television was invented, film narratives developed several techniques for indicating simultaneity: titles such as "meanwhile, back at the ranch," placing large clocks in every location, verbal indicators in the dialogue, parallel montage, etcetera. My point is that each television narrator not only has to choose which indicator of simultaneity to employ but also make decisions as to which simultaneous story event to show first, which next, and how long to stay with each scene before cutting back. One set of choices will create tension and suspense, another will be lackadaisical.

Not only can discourse reorder the sequence of story events, it can also alter events' duration. Building on Gerard Genette's work in *Narrative Discourse,* Seymour Chatman details the following five possible matches between story and discourse duration:[19]

1) Summary: Discourse-time is shorter than story-time.

Verbal narratives rely heavily on summarizing story events. A narrator who remarks, "I went to college for three years and then dropped out to become a woodworker," has compressed events that took several years into one brief sentence. In visual media, summary is less common and slightly awkward because time condensation is more difficult without verbal tenses. (Forties films tend to use such clichés as calendar leaves rustling in the wind.) Perhaps the closest that television comes to summary is in montage sequences (particularly those used in tandem with voice-over narration). Thus the title sequence of *Gilligan's Island* condenses events that must have taken some hours or days into a minute or so.

2) Ellipsis: The same as summary except that discourse-time is zero.

Television narratives depend upon ellipsis. Every time the camera cuts from a person leaving a building to that same person getting out of his car, it has cut out all the story-time in between. This habit of eliding out routine events or nonpertinent stretches of time allows television to present a story that supposedly has a duration of several hours, days, weeks, or months within the confines of a half-hour or hour discourse.

3) Scene: Story-time and discourse-time are equal.

Whenever a television show allows the camera to present story events in full, without temporal cuts (the camera can change its spatial position at will so long as no time is lost), we have congruence between story and discourse-time. The scene is the basic building block of television narratives, the mode of choice for all conversations and confrontation.

4) Stretch: Discourse-time is longer than story-time.

The best example of stretch is slow motion. In slow motion the narrator takes longer to relate the events than the events originally lasted in the story. (Fast motion, which is rarely used, would qualify as a form of summary.)

5) Pause: The same as stretch except that story-time is zero.

One example of a pause would be freezing the frame completely while the sports announcer analyzes the action. Commercials also use pauses; at the end of the Nyquil commercial mentioned earlier, we get a freeze frame of the couple in bed and a superimposed picture of the product, while print and voice-over simultaneously proclaim: "Vicks Nyquil, the nighttime sniffling, sneezing, coughing, aching, stuffy head, fever, so you can rest medicine, from Vicks, of course." The action of the story has paused, but the narrator continues to speak and to drive home the moral of the story.

As Genette also pointed out, narratives have several options in terms of their correspondence between story and discourse frequency. Each narrator has a choice between the "singulative," the "repetitive," and the "iterative." That is to say, a narrator can: tell once what happened once (one shot of the quarterback's brilliant pass), tell n times what happened once (replaying the shot of the pass n times), or tell once what happened n times (using one shot of one brilliant pass to stand for all the brilliant passing the quarterback did in that game).

What is the point of identifying these time distortions? For one thing, it can be intriguing to consider what lies behind the temporal choices. Interestingly enough, commercials often strategically elide story-time; they cut from the "before" situation to the "after"—we see the dirty shirt and then the clean one—but all the work of doing the laundry is lost in the shuffle. Similarly, a show may begin with some exciting action to grab the viewer's interest and only flash back to provide less eye-catching background information once its hold on the viewer is firmly established.

Furthermore, examining the temporal distortions can help us characterize television narrators. The closer the discourse approaches to congruence with story-time through presenting singulative scenes in chronological order, the less interventionist and the more invisible the narrator; the more the discourse distorts story-time through achronological order, unusual pacing, or repetition, the more the narrator's hand is revealed. Sitcoms tend to have self-effacing narrators; they proceed chronologically from scene to scene, whereas rock videos make strange time distortions part of their style.

Moreover, narrative theory provides us with a framework for understanding one of the unique qualities of television—the ability to broadcast "live." "Liveness" may be defined by the congruence of discourse-time and reception-time, that is, no time gap exists between the narrative's production and its consumption. This quality of "liveness" is rare in our contemporary experience of narratives (we are used to films and novels having been "spoken" many years before we happen

to screen the movie or buy the book), and thus rather exciting and inti-
mate. In a sense, "liveness" is a throwback to traditional oral storytelling, in which, after all, there would be no temporal gap between the
speaker speaking and the listener listening. (Yet there is an important
difference between oral and television "liveness": with TV the viewer
can seldom be sure of the simultaneity of discourse production and
reception.)

There are two subsets of "live" television broadcasts. The first is
comprised of those shows in which only the discourse is live. On the
CBS Evening News, depending on where you live and when the local
station decides to run the national news, Dan Rather might really be
speaking at the moment you are listening to him, but the news that he
or the other correspondents relate still took place earlier in the day, so a
distinct gap still exists between story-time and discourse-time. On certain other occasions, such as the coverage of sporting events or catas-
trophes, we will hear the narrator/sportscaster/reporter narrating at
the same moment that the events are transpiring. In such cases, both
story and discourse are "live." However, one might note that such special circumstances are characteristically brief; generally, the sports-
caster or reporter will soon interrupt the transpiring events for commentary, flashbacks, recaps, etcetera. Thus the moments of temporal
convergence between story and discourse are little islands in a complicated, shifting temporal relationship that foregrounds the narrator's
control.

In certain cases, however, such as the initial broadcasts of dramatic
shows of the fifties or the skits on *Saturday Night Live,* the stories
are allowed to transpire before us without overt narratorial control
and without interruption or temporal disfigurement. In such cases,
reception-time equals discourse-time equals story-time. I would argue
that these shows reach a limit point at which narrative almost shades
off into drama. True, live television plays differ from theatrical plays in
that the multicamera setup still determines our view of the proceedings and in that the actors are not truly "present" before the audience.
Yet the sustained convergence between the three time schemes and
the absence of voice-over commentary effectively downplays the presence of a narrator and creates a powerful sense of theatrical immediacy.

With television narratives, there is not only the time of the told and
the time of the telling, but also the time of the broadcasting. Let us
turn now to look at this third, outermost layer.

SCHEDULE

Short stories, novels, and films can all be seen as "freestanding"; the reader or viewer has relatively unfettered access to these kinds of narratives, and the narratives themselves are allowed to develop without too many blatant outside constraints. Television narratives are unique in the fact that each narrative is itself embedded within the meta-discourse of the station's schedule. A viewer can circumvent some of the extrinsic consequences of this embedding by using a videocassette recorder; one can, for example, watch a text at a more convenient time, or watch it more than once, or fast-forward through interruptions. But this embedding has also forced television narratives to make certain intrinsic adjustments of story and/or discourse. In this section we shall look at a few of the most common adjustments.

American television schedules are like jigsaw puzzles. They are composed of scores of separate pieces that must fit together in certain patterns and thus must conform to standardized rules. For openers, each piece of the jigsaw puzzle must fit into a neat time frame—thirty seconds, a minute, thirty minutes, sixty minutes, whatever. Accordingly—unlike oral, literary, or film narratives, which are much more likely to last as long as their story requires—television narratives have to fit their assigned Procrustean bed. This frequently means that two-hour television movies and miniseries are "padded" with insignificant events, whereas certain commercials and news stories don't have enough time to develop their stories before they must conclude.

Another principle of most television schedules is that each text must accommodate interruption. The most common form of interruption, of course, is the commercial break, but one should not overlook the "pledge breaks" on public television stations nor the "kitchen and bathroom" intermissions that cable networks insert into long feature films. Television narratives have learned to compensate for and even take advantage of the inevitable interruptions in various ways. First, they typically tailor their discourse to fit "naturally" around the commercial breaks, so that, for instance, the exposition fits before the first break and the coda after the last. (In "Old Acquaintance," the first segment presents the theft of the dolphin, the last "rhymes" by presenting the dolphin returned to its aquarium and Goldie's farewell.) Second, shows build their stories to a high point of interest before each break to insure that the audience will stay tuned. Finally, shows frequently time the placement of commercials to coincide with a temporal ellipsis so that while the viewer's attention has been diverted, the story can gracefully leap ahead several hours or days.

Because most television stations broadcast either round the clock or nearly so, they have a voracious demand for material. To maximize investments in time and money, it is cheaper to continue using the same cast and set than to create all new shows. Moreover, as writers of comic strips, popular novels, and radio shows had already discovered, using the same existents has the advantage of building audience familiarity and loyalty. Thus, as we all know, few television narratives are self-contained, single broadcasts; a large proportion continue each day or each week. Continuing shows can be divided into two categories. "Series" refers to those shows whose characters and setting are recycled, but whose story and discourse conclude in each individual sitting; a "series" is thus similar to an anthology of short stories. By contrast, in "serials," both the story and discourse continue after a day or a week's hiatus; they are similar to serialized Victorian novels. (Note that according to this definition, a series can temporarily become a serial by offering a "two-parter.") Serials can be further divided into those that do eventually end (despite the misnomer, I would place miniseries in this category), and those, such as soap operas, that may be cancelled but never reach a true conclusion.

We have already examined many of the consequences of the series format when we looked at television stories. For one thing, because the characters must continue from week to week, suspense is minimized; the viewer knows that Magnum is never in great danger. For another, the series format requires a constant stream of new antagonists and new love interests. Moreover, series characters all seem struck with selective amnesia. It is often said that series have no memory and no history; this is partially true in that none of the characters seem to notice that they did exactly the same things the previous week. And yet the characters' problems and relationships do not continually start over from scratch; while past events seem to disappear into a black void, characters do build their relationships from week to week—one can note, for instance, a progression in Mary Richards's relationship with Lou Grant. Still, as long as the series continues, the viewer can bank on the fact that the central tension or premise will not be resolved: Magnum will never "grow up," the *Enterprise* will never complete its mission and return to earth. Only on rare and red-letter occasions will a series resolve its central premises: the last episode of *M*A*S*H* created such a stir not only because it was the last episode of a popular series but because the show actually created a new state of affairs—the Korean war ended and everyone got to go home.

As for serials, one of the problems they face is how to bring up to date viewers who do not usually watch the show or who have missed an episode. To this end most serials begin with a flashback recap of ongoing storylines. Another option, used most frequently by daytime soap operas, is to have the characters redundantly discuss the most significant past events. Similarly, serials have to generate enough viewer interest and involvement to survive their hiatus. Some offer flashforwards to tease the viewer with bits of upcoming action. Frequently, they also turn to the technique made famous by movie serials—the cliffhanger. Thus, a *Hill Street Blues* episode will end with Bobby Hill and Andy Renko getting shot, a *St. Elsewhere* episode will conclude with an antiabortionist's bomb ominously ticking away undiscovered. In a sense the suspense that these endings create is a trick; one knows that it is quite unlikely that *Hill Street Blues* could afford to kill off both Hill and Renko, but the cliffhanger leaves the viewer a little uneasy on the subject just the same. Moreover, there seems to be an unwritten rule that the longer the hiatus, the higher the cliff; witness the spectacular cliffhangers whipped up on *Dallas* and *Dynasty* for the last show of the season.

Because of its hunger for material, the television schedule also relies on out-and-out repetition. Television texts are played again and again: ongoing series repeat themselves each spring and summer season; older series are replayed endlessly in syndication; once "live" programs reappear as "canned"; cable and PBS networks shamelessly play the same program or film over and over in different time slots. As discussed before in reference to commercials, one effect of this method of presentation on television narratives is that they have to find ways to catch the interest of a viewer who might have seen them before. Another effect may be that these narratives will studiously avoid topical issues, because what is topical when the show is produced will be out of date when the series is in syndication.

One of the questions asked at the chapter's opening was "What qualities are specific to television narratives?" Because television offers so many disparate types of narrative texts (commercials obviously are quite different from soap operas or made-for-TV-movies), it is difficult to generalize. Similarly, many of the qualities one perceives as characteristic of television can also be found on radio or in serialized novels. Keeping firmly in mind, then, that these characteristics are neither exclusive nor omnipresent, one could hazard the following list of television narrative's most common traits: series or serial format; predictable, formulaic storylines; multiple storylines intertwined in complex patterns and frequently interconnecting; individualized, appealing characters fitting into standardized roles; emphasis on the interrelationships between these characters; endings of texts mark a return to the same state of affairs; settings and scenery either very showy or merely functional; substitute narratees; voice-over narration, and direct address often employed; most narrators omniscient and reliable; reliance on ellipsis and scene; achronological order to entice (previews) or inform (flashbacks); accommodation of interruptions; lengths cut to fit standardized time slots; and a tendency towards universality, away from topicality.

I have been treating the television schedule as a kind of discourse. In a sense, I believe that we can also look behind each station's schedule to see a kind of supernarrator. These supernarrators are personified and individualized by various means: logos (the NBC peacock, the CBS eye); signature music; and, most importantly, voice-over narrators who speak for the station or network as a whole. Indeed, each station routinely uses certain voice-over narrators who speak to the viewer, providing flashforwards of coming attractions, justifying schedule changes or technical difficulties, interrupting for news bulletins or tests of the emergency broadcast system. (Nearly all of these voice-over narrators are male—to my knowledge only PBS ever uses a female spokesperson.)

Because they are the narrators of the outermost frame, these strange storytellers are in the position of the most power and knowledge. They sit outside and above all the embedded narratives, unaffected by them. One has very little information by which to judge their objectivity, sincerity or reliability. And it is through their sufferance that all the other texts are brought to us; they can interrupt, delay, or preempt the other texts at will.

And yet, these network narrators are hardly frightening, because they have an Achilles heel. They have a desperate need—viewers. "Stay tuned, stay tuned, stay tuned," they entice and entreat. They seem haunted by a nightmarish image—suppose they told a story and nobody came? We viewers hold the ultimate power, to listen and allow them to exist, or to plunge the screen into darkness.

NOTES

1. Shlomith Rimmon-Kenan, *Narrative Fiction: Contemporary Poetics* (London: Methuen, 1983), p. 15.

2. Tzvetan Todorov, "The Grammar of Narratives," in *The Poetics of Prose*, trans. Richard Howard (Ithaca, N.Y.: Cornell University Press, 1977), p. 111.

3. Roland Barthes, "Introduction to the Structural Analysis of Narratives," in *Image/Music/Text*, trans. Stephen Heath (New York: Hill and Wang, 1977), p. 93.

4. Aristotle, "Poetics," in *Critical Theory since Plato*, ed. Hazard Adams (New York: Harcourt, Brace, Jovanovich, 1971), p. 52.

5. Gustav Freytag, *Freytag's Technique of the Drama: An Exposition of Dramatic Composition and Art*, trans. Elias MacEwan (New York: Benjamin Blom, 1968), pp. 114–40.

6. Vladimir Propp, *Morphology of the Folktale* (Austin: University of Texas Press, 1970), pp. 19–20.

7. Ibid., pp. 21, 22, 23.

8. As outlined by Robert Scholes, *Structuralism in Literature: An Introduction* (New Haven, Conn.: Yale University Press, 1974), pp. 63–64.

9. Peter Wollen, "*North by Northwest*: A Morphological Analysis," *Film Form*, no. 1 (1976), pp. 19–34.

10. For descriptions of other "narrative grammars" see Rimmon-Kenan, *Narrative Fiction*, pp. 22–28, 34–35; Scholes, *Structuralism in Literature*, pp. 95–117.

11. Roland Barthes, *S/Z*, trans. Richard Miller (New York: Hill and Wang, 1974).

12. Robert C. Allen, "On Reading Soaps: A Semiotic Primer," in *Regarding Television—Critical Approaches: An Anthology*, ed. E. Ann Kaplan, American Film Institute Monograph Series, vol. 2 (Frederick, Md.: University Publications of America, 1983), p. 103.

13. David Marc, *Demographic Vistas: Television in American Culture* (Philadelphia: University of Pennsylvania Press, 1984), p. 12.

14. Robert Scholes and Robert Kellogg, *The Nature of Narrative* (New York: Oxford University Press, 1966), pp. 240–82.

15. Christian Metz, "Notes Toward a Phenomenology of the Narrative," in *Film Language: A Semiotics of the Cinema*, trans. Michael Taylor (New York: Oxford University Press, 1974), p. 21.

16. John Fiske and John Hartley, *Reading Television* (London: Methuen, 1978), pp. 85–100.

17. Walter Benjamin, "The Storyteller," in *Illuminations*, ed. Hannah Arendt, trans. Harry Zohn (New York: Schocken Books, 1968), p. 92.

18. Metz, "Phenomenology of the Narrative," p. 18.

19. Seymour Chatman, *Story and Discourse: Narrative Structure in Fiction and Film* (Ithaca, N.Y.: Cornell University Press, 1978), p. 68. See also Gerard Genette, *Narrative Discourse: An Essay in Method*, trans. J. E. Lewin (Ithaca, N.Y.: Cornell University Press, 1980), pp. 33–160.

FOR FURTHER READING

General overviews of narrative theory can be found in: Seymour Chatman, *Story and Discourse: Narrative Structure in Fiction and Film* (Ithaca, N.Y.: Cornell University Press, 1978)—highly readable, it basically deals with literature but pays some attention to film; Wallace Martin, *Recent Theories of Narrative* (Ithaca, N.Y.: Cornell University Press, 1986)—compares competing theories, includes useful diagrams and examples, and offers a thorough bibliography; Shlomith Rimmon-Kenan, *Narrative Fiction: Contemporary Poetics* (London: Methuen, 1983)—both concise and thorough, it contains helpful contrasts between theorists' works and has an excellent annotated bibliography.

Highly recommended studies of specific aspects of narrative are: Roland Barthes, "Introduction to the Structural Analysis of Narratives," in *Image/Music/Text*, trans. Stephen Heath (New York: Hill and Wang, 1977)—pioneer work that quickly covers a great deal of ground and is particularly influential for its treatment of story events; Wayne C. Booth, *The Rhetoric of Fiction* (Chicago: University of Chicago Press, 1961)—a classic for all students of literature, it covers point-of-view issues with keen intelligence; Jonathan Culler, *Structuralist Poetics* (Ithaca, N.Y.: Cornell University Press, 1975)—best illustration of narrative theory's debt to structuralism, and important for its coverage of "naturalization" and convention; Gerard Genette, *Narrative Discourse: An Essay in Method*, trans. J. E. Lewin (Ithaca, N.Y.: Cornell University Press, 1980)—a sustained analysis of Marcel Proust's *A la recherche du temps perdu* mixed with rigorous theory, his discussion of time and point of view in narrative have served as models for all subsequent work; Christian Metz, "Notes Toward a Phenomenology of the Narrative," in *Film Language: A Semiotics of the Cinema*, trans. Michael Taylor (New York: Oxford University Press, 1974)—a brief essay that offers a thorough and suggestive definition of narrative; Vladimir Propp, *Morphology of the Folktale* (Austin: University of Texas Press, 1970)—pathbreaking study of story structure that is short, readable, intriguing; Robert Scholes and Robert Kellogg, *The Nature of Narrative* (New York: Oxford University Press, 1966)—a historical overview of narrative form from the Greeks to the twentieth century, it is theoretically outmoded, but an important antidote to ahistorical theorizing.

To date, narrative theory's applicability to television has not been thoroughly studied. The following texts, however, can be quite useful: Robert C. Allen,

Speaking of Soap Operas (Chapel Hill: University of North Carolina Press, 1985)—applies narrative theory to soap operas; David Bordwell, *Narration in the Fiction Film* (Madison: University of Wisconsin Press, 1985)—lengthy and scholarly discussion drawing on the work of Genette, Chatman, and the Russian Formalists; John Ellis, *Visible Fictions: Cinema, Television, Video* (London: Routledge and Kegan Paul, 1982)—compares film and television as narratives, but without any particular reliance on narrative theory; E. Ann Kaplan, ed., *Regarding Television—Critical Approaches: An Anthology*, American Film Institute Monograph Series, vol. 2 (Frederick, Md.: University Publications of America, 1983)—contains essays by Robert Stam, Sandy Flitterman, Robert C. Allen, and Maureen Turim that discuss different types of television shows as narrative; Bruce Kawin, *Mindscreen: Bergman, Godard, and First-Person Film* (Princeton, N.J.: Princeton University Press, 1978)—first half of the book provides useful discussion of film's narrativity, particularly the difference between primary and embedded narrators; Sarah Kozloff, *Invisible Storytellers: Voice-Over Narration in American Fiction Film* (Berkeley: University of California Press, forthcoming)—uses narrative theory in discussing cinematic voice-over narration, but most of the discussion is applicable to television as well; David Marc, *Demographic Vistas: Television in American Culture* (Philadelphia: University of Pennsylvania Press, 1984)—good general discussion of the characteristics of television stories.

READER-ORIENTED CRITICISM AND TELEVISION
ROBERT C. ALLEN

"Reader-response criticism," "reception theory," and "reader-oriented criticism" are all names given to the variety of recent works in literary studies that foreground the role of the reader in understanding and deriving pleasure from literary texts. Traditionally, says Wolfgang Iser, a leading force in the German variant of reader-oriented studies, critics have regarded the literary text as something that possesses meaning much in the way that an oriental carpet possesses a pattern. Thus critics saw their task as finding the "figure in the carpet"—the meaning of the work that lay hidden in its structure—and relating that meaning to other readers who had not discovered it for themselves (or who did not possess the interpretative gifts of the critic).

To Iser, the presumptions of traditional literary criticism render the entire critical enterprise little more than intellectual strip mining: the critic plows through the text looking for signs of hidden meaning. When all the bits of meaning have been extracted from the textual site, the critic displays them before a suitably grateful public and then moves on to mine another text. Iser sees such an approach to criticism as fatal both for literature and for criticism, because if meaning is something that can be extracted from a text like coal from a hillside, then the act of criticism reduces literature to a pile of "used-up" texts about which there is nothing left to be said.[1]

Iser's attack on this "archeological approach" to criticism is directed at its most basic underlying assumption: that meaning is, to use Thomas Carlyle's phrase, an "open secret" waiting to be found by the insightful critic and reader. Whether the critic then ascribes that found meaning to the intention of the author, the author's unconscious motives, the "spirit" of an age, or the relationships among formal elements in the text, meaning nevertheless is seen to exist "out there," in the text, independent of the mind of the critic or reader.

All of the critics whose work I am lumping together under the category of reader-oriented criticism share with Iser the fundamental belief that this schema ignores a crucial fact of literature—namely, that works are made to mean through the process of reading. Although they disagree about many other things, literary critics and theorists as diverse as Iser, Roman Ingarden, Georges Poulet, Mikel Dufrenne, Hans Robert Jauss, Stanley Fish, Norman Holland, Jonathan Culler, and Tony Bennett would agree that literary meaning should no longer be viewed as an immutable property of a text but must be considered as the result of the confrontation between reading act and textual structure. In other words, for reader-oriented critics, what previous critics took for granted has become the central focus of critical investigation: what happens when we read a fictional narrative?

The common-sense observation that meaning does not occur except through the reading act has given rise not so much to a single approach to literature (and by extension to film, television, and other forms of cultural production) as to a field of inquiry. Critics and theorists I would call reader-oriented sometimes share little more than a common starting point for their projects—an insistence upon admitting the reading act to the critical agenda. The questions that logically flow from this starting point have hardly been answered in a single voice, perhaps because they come so close to the heart of criticism and to our relationship with those curious other worlds we call literature, or film, or television. Rather than try to disentangle the many critical skeins that go into reader-oriented criticism as a prelude to demonstrating how they might be applied to television, I will organize my discussion of this approach around what I see to be a set of key questions—keys both to the project of reader-oriented approaches in general and to their possible application to television narratives. In doing so I will necessarily emphasize the work of a few critics over that of others and gloss over philosophical and methodological differences that the proponents of the various "schools" of reader-oriented criticism delight in calling to each other's attention.[2] What is most important to the student of television is the phenomenon that reader-oriented criticism thrusts into the critical foreground—the reading act—and the questions, issues, and opportunities that arise when the text being read is televisual rather than literary or cinematic.

HOW DO WE READ A
FICTIONAL NARRATIVE TEXT?

The organization of reader-oriented criticism around this central question might suggest that literary theorists did not concern themselves with it before the emergence of reader-oriented criticism in the 1960s. Like all critical approaches, however, reader-oriented criticism did not emerge from a philosophical vacuum. Despite the near irrelevance of this question for the "figure-in-the-carpet" tradition of criticism, it formed one focus of study for scholars within the branch of philosophy and literary theory called *phenomenology* decades before the publication of works by Iser and Jauss in the 1960s and 1970s.

Given its name by philosopher Edmund Husserl in the 1930s, phenomenology concerns itself with the relationship between the perceiving individual and the world of things, people, and actions that might be perceived. These are not two separate realms connected only by the passive sensory mechanisms of the individual, declared Husserl, but rather they are inextricably linked aspects of the process by which we know anything. All thought and perception involve mutually dependent subjects and objects. I cannot think but that I think *of something*. Thus to study any *thing* is to study that thing as it is experienced or conceptualized within the consciousness of a particular individual. Reality, in other words, has no meaning for us except as individually experienced phenomena.

Phenomenology provides the philosophical basis for the work of a number of literary theorists who have influenced reader-oriented criticism—Hans-Georg Gadamer, Roman Ingarden, the so-called Geneva School of criticism of the 1930s and 1940s—and directly informs much of contemporary reader-oriented criticism itself. For the above-mentioned scholars, reading the fictional narrative text provides an especially interesting case of the more general process by which subjects (individuals) take objects into their consciousness or "intend" them, to use Husserl's term. By "intend" Husserl means not so much "to want it to be like" (as in, "He intended that things should go well") as "to direct one's conscious sensory and sense-making capacities toward."

The phenomenologist's fascination with the act of reading lies in the curious and paradoxical process by which lifeless and pitifully inadequate words on a page are not just made to mean something through the intentions of the reader but are "brought to life" in that reader's imagination. This process occurs in reading the simplest fictional narrative (a joke, folktale, or anecdote) and in the most complex literary experience (slogging through *Finnegan's Wake,* for example). It occurs so quickly and so automatically that it would appear to short-circuit conscious logic. The world constructed as a result of the reading act has existence only in the mind of the reader, and yet its construction is initiated and guided by words that exist "out there" on the page. Furthermore, those words on the page were "intended" (in both senses of the term) by another consciousness—that of the author who wrote them. Geneva School critic Georges Poulet, for example, describes the reading act as an acquiescence by the reader to the thoughts of another consciousness. "Because of the strange invasion of my person by the thoughts of another, I am a self who is granted the experience of thinking thoughts foreign to him. I am the subject of thoughts other than my own."[3]

Obviously, the relationship between text and reader can be conceptualized in a number of ways—as a sort of mutually sustaining collaboration, a surrender to the thoughts of another, or even a battle of wills between the intentions of the reader and those of the author. The analogies differ according to the degree of determinacy each critic assigns the reader's intentional activity, the text, and the author's intentions. But let us take one conceptualization of the reading act and follow it through from its origins in phenomenological theory to its application to literature and television.

For Roman Ingarden, a student of Edmund Husserl, the literary text starts as an intentional act on the part of the author. Once the work has been written and published, however, it exists separately from those originating intentions. The analogy Ingarden uses is that of a musical composition. The musical text certainly has a material status as written notes on paper that have resulted from the composer's activity. At this point, however, the text is still only a set of possibilities. The musical text becomes a musical work only when a performer "concretizes" the text in performance. The text exists apart from any particular rendering of it, but the work has meaning for us only as a performance. Similarly the literary text for Ingarden is but a "schemata," a skeletal structure of meaning possibilities awaiting concretization by the reader's own intentional activity. In a very real sense, then, each reading is a performance of meaning.[4]

As words on the page, the literary text is but one-half of the perceptual dynamic; it is an object, yet without a perceiving subject. Or, seen another way, it is the material residue of an absent, intending subject—the author. In the reading act, the fictional world represented by the words on the page is rendered within the consciousness of the reader. That world is created as the reader follows the directions for meaning construction provided by the text, but even more importantly as the reader fills in the places the text leaves vacant.

This notion of reading as "gap filling" is extremely important to Ingarden and, indeed, to the entire project of reader-oriented criticism. It is an extension of the more general phenomenological theory of perception. In making sense of the world around us we intend things, concepts, actions—that is, we impose some sense of meaning upon experiences by directing our conscious faculties toward them. When the objects of our perceptions (and hence intentions) are "real" objects, the meanings we attach to them are to some degree limited by their material qualities. The cat sitting on my lap at the moment is not just any cat, but *this* small, purring, four-year-old, female tabby named Dorothy. Still, however, I can only perceive a few of the manifold aspects of this particular cat at any given moment. Looking down at her now, I see her as a rather indistinctly shaped mass of fur with no face, feet, or tail. I can pick her up and thus reveal these hidden aspects, but in doing so I necessarily obscure others. Nothing can be perceived in all its aspects at once, and we constantly make sense of the world by extrapolating from a small number of qualities to a whole thing.

The cats we experience in literary narratives, however, have no existence in the "real" world of things. They exist only as words on a page and thus are much more indeterminate than real cats. If, for example, I read in a story: "The man sat at his computer terminal with a sleeping cat on his lap," I as the reader of this tale can imagine all sorts of qualities this fictional cat might possess. The text, as it were, leaves it up to my intentions to specify whether it is male or female; black, yellow, or multicolored; friendly or temperamental; small or large; spayed or not; and so on. In fact, the text leaves it up to me to decide how determinate a cat I want to make it at all. I can quite legitimately construct or intend whatever kind of cat I like so long as my intentions do not contradict information the text gives me. No matter how elaborately the text describes this cat, the construction of it by the reader will always be a process of making a whole cat out of an incomplete set of cat descriptors by supplying the missing parts.

One of the amazing things about the worlds we construct in reading literature is that those worlds appear to us to be fully formed and complete from the time we get our first descriptions of them on page one until after we have finished reading the final paragraph of the book. Reading a novel is not like playing "connect the dots." We don't start with an apparently random arrangement of words that take on meaning and life only at the end of the reading process. To phenomenologists this experiencing of a narrative world as fully formed from the beginning provides evidence for the crucial role of intention in our negotiation of worlds, both textual and material. Even on the basis of the tiniest scrap of information, we will provide whatever is missing until we have organized our perceptual field into objects that make sense to us.

Gap filling is also affected by our movement through the text. The confrontation between our initial expectations and the text forms a sort of provisional fictional world, on the basis of which we develop further expectations of what is likely to happen next, as well as assumptions about the relationship between any one part of this fictional world and any other. As we read further, those expectations are modified in order that we can keep a coherent world before our mind's eye at all times. Furthermore, the text keeps shifting our perspective on this world—foregrounding this aspect in one chapter; that one later on. In short, Ingarden reminds us that reading is a dynamic tension between the reader's expectations and the text's schematic instructions for meaning production. The result is a constantly changing fictional world, but one that appears to us as whole and complete at any given moment during the reading act.

Ingarden's description of the reading act becomes the starting point for Wolfgang Iser's *The Act of Reading*.[5] Iser points out that our relationship with narrative artworks is fundamentally different from that with painting or photography. A painting is available to us all at once. The only time we experience a novel or film as a whole, however, is when we have finished experiencing it—that is, when we are no longer reading it. Instead of being outside the work contemplating it as a whole, the reader of a narrative takes on what Iser calls a "wandering viewpoint": a constantly changing position within the text itself. Iser calls this relationship between dynamic reading activity and unfolding textual terrain "unique to literature." The photographic basis of the cinema (and, we may assume, television) gives it too much of an "all-at-onceness" at the level of the shot for Iser's taste. It is clear, however, that any narrative form involves the reader's (or viewer's) movement through the text, from one sentence, shot, or scene to the next.

As the term "wandering viewpoint" suggests, Iser's theory of the reading act emphasizes its diachronic (occurring over time) dimension. At every moment during the reading act, we are poised between the textual geography we have already wandered across and that we have yet to cover. This tension, between that which we have learned and that which we anticipate, occurs throughout the text and at every level of its organization. Each sentence of a literary narrative both answers questions and asks new ones.[6] Iser describes this process as an alternation between protension (expectation or anticipation) and retention (our memory of the text to that point). Each sentence does not so much fulfill our expectations as it alters and channels them. To continue the geographic metaphor, each new "block" of text we cover provides us with a new vantage point from which to regard the landscape of the text thus far, while, at the same time, it causes us to speculate as to what lies around the next textual corner. Hence our viewpoint constantly "wanders" backwards and forwards across the text.

According to Iser, although the text can stimulate and to some degree channel protension and retention, it cannot control those processes. This is because protension and retention occur in the places where the text is silent—in the inevitable gaps between sentences, paragraphs, and chapters. It is in these holes in the textual structure that we as readers "work" on that structure. We make the connections that the text cannot make for us.

Iser's theory of reading activity as gap filling relies upon a basic semiotic distinction—discussed in Ellen Seiter's essay—between paradigmatic (associative) and syntagmatic (sequential) organization. (Perhaps the easiest way to keep straight the difference between paradigmatic and syntagmatic axes of textual organization is to think of them as the principles by which a restaurant menu is organized. Items are arranged paradigmatically, with all those having a basic affinity grouped together: appetizers with appetizers, entrées with entrées, desserts with desserts. The menu is also arranged syntagmatically according to the sequence of dishes in an ordinary meal: appetizers appear first, then soups, then fish dishes, then entrées, then desserts, etc.) The gaps Iser speaks of in the text involve the syntagmatic arrangement of textual segments—the space between one chapter and the next, for example. These gaps provide us with an opportunity to consider possible paradigmatic relationships between them as well—how might one be related to the other conceptually?

Underlying Iser's account of the reading activity is the notion of "consistency-building," which derives from Ingarden and from phenomenology in general. The connections that readers make between textual segments are those that contribute to the maintenance of a coherent textual world. Again we have the idea that, faced with an ambiguous, unconnected, or even seemingly random set of sensory experiences, we will impose some sort of coherence and order upon them. Because the fictional narrative gives us no material points of reference by which to order its world, we are continually adjusting our picture of it to fit with new information the text presents us. On the other hand, we tend to foreground that new information that fits most easily into our existing view of the text's world (our *Gestalt*) and leave on the periphery of our attention elements that cannot immediately be correlated with that which we already know. It is at the level of our totalizing view of the text's world, its *Gestalt*, that a text has meaning for us. And because this *Gestalt* is imposed by the reader upon the structure of the text, it is only in the mind of the reader that a text becomes a work of literature.[7]

It is obvious from the above discussion that Iser limits his theory of the reading act to literature. In fact, as I have suggested, he seems to regard the process by which we understand films and television as inherently different. We should also note that, although the fundamentals of the reading process Iser outlines should be applicable to the experience of reading any type of fictional narrative, Iser's examples are drawn almost exclusively from "high-art" literature—*Pilgrim's Progress, Ulysses, Tristram Shandy*. Thus, Iser himself might be horrified at the prospect of someone applying his theory of reading not just to television, but to one of the most popular and least "artsy" of television narrative forms—the soap opera. Yet this is precisely what I propose to do. I believe that, regardless of the range of texts to which it was intended to be applied, the phenomenological theory of reading activity developed by Ingarden and elaborated by Iser helps to account for the relationship between soap opera viewers and the curiously structured and quite complex fictional worlds they encounter daily. (I am tempted to say it provides a guiding light into another world, but I won't.) Furthermore, given that some aspects of "reading" soap operas overlap with the processes involved in reading any narrative broadcast on commercial television, a reader-oriented account of the relationship between soap operas and their viewers might help us to understand our relationship with television narratives more generally.

Several things immediately strike us about the soap opera as a narrative structure. The first is the staggeringly large amount of text devoted, ostensibly at least, to the relating of the same story. Each year an hour-long soap opera offers its viewers 260 hours of text. Most of the soap operas currently being run on American commercial television have been on the air for at least ten years. In cinematic terms, this represents the equivalent of 1,300 feature-length films! Two soap operas, *Guiding Light* and *Search for Tomorrow,* have enjoyed continuous television runs since the early 1950s, giving them each texts that would take more than a year of nonstop viewing to "read."

Another distinctive feature of the soap opera text is its presumption of its own immortality. Although individual subplots are brought to temporary resolution, there is no point of final narrative closure toward which the soap opera narrative moves. Individual episodes advance the subplots incrementally, but no one watches a soap opera with the expectation that one day all of the conflicts and narrative entanglements will be resolved so that the entire population of Port Charles or Pine Valley can fade into happily-ever-after oblivion.

A final resolution to a soap opera's narrative seems so unlikely in part because we follow the activities of an entire community of characters rather than the fate of a few protagonists. It is not at all unusual for a soap opera to feature more than forty regularly appearing characters at any given time—not including those characters who have been consigned to the netherworld between full citizenship in the community and death: characters who are living in London, New York, or some other distant city; or whose fate is "uncertain." These large communities represent elaborate networks of character relationships, where "who" someone is is a matter of to whom he or she is related by marriage, kinship, or friendship. These complex character networks in daytime soap operas distinguish them even from their prime-time counterparts such as *Dallas* and *Dynasty.* Although the latter are serial narratives (like daytime soaps) and foreground character relationships, their networks are much smaller and are organized around a few central characters. The community of a nighttime soap consists of only twelve to fifteen regular characters at any given time.

In an attempt to account for the soap opera viewing process, we might begin by recalling Iser's point that we can never experience a narrative work in its totality while we are reading it; we are always someplace "inside" its structure rather than outside of it contemplating it as a whole. However, unlike closed narrative forms (the novel, the short story, the feature film, the made-for-TV movie), the soap opera does not give us a position after "The End" from which to look back on the entire text. The final page of a soap opera never comes, nor is it ever anticipated by the viewer. As soap opera viewers, we cannot help but be inside the narrative flow of the soap opera text. Furthermore, our "wandering" through the soap opera text as viewers is a process that can occur quite literally over the course of decades.

Even if we wished to view the entire text of *All My Children* or *General Hospital* to this point in its history we would be unable to do so. Neither could we view portions of previous textual material that we missed or wished to re-view (unless, of course, we have been videotaping each episode and saving them). Our viewing of soap operas is regulated by their being parceled out in weekday installments. Certainly, it is a characteristic of films and television programs that, unlike literature, the rate at which we "read" is a function of the text itself rather than our reading activity. Except where we manipulate the "special effects" features on videotape recorders, the images on the screen flash by at a predetermined and unalterable rate. With soap operas, and to a lesser degree with other series and serial forms of television narrative, this reading regulation is not just technological, but institutional as well—a measured portion of text is allocated for each episode and for each scene within an episode. Unlike the series form of television narrative, where a complete story is told in each episode and only the setting and characters carry through from week to week, the soap opera simply suspends the telling of its stories at the end of the hour or half-hour without any pretext of narrative resolution within a given episode. Unlike the radio soap operas of the 1930s and 1940s, the television soaps of today do not end each episode with an announcer's voice asking: "Will Mary forgive John's thoughtlessness and agree to marry him? Join us tomorrow. . . ." However, the calculated suspension of the text at the end of each episode of a television soap implicitly encourages the viewer to ask the same sort of question and provides the same answer: you'll have to tune in tomorrow to find out.

Viewed in terms of reader-oriented criticism, the time between the end of one soap opera episode and the beginning of the next constitutes an institutionally mandated gap between syntagmatic segments of the text. Iser comments on a parallel pattern of textual organization in the novels of Dickens and in serialized fiction in magazines. During Dickens's lifetime most of his readers read his novels in weekly magazine installments, rather than as chapters of a single book. In fact, says Iser, they frequently reported enjoying the serialized version of *The Old Curiosity Shop* or *Martin Chuzzlewit* more than the same work as a book. Their heightened enjoyment was a result of the protensive tension occasioned by every textual gap (What's going to happen next?) being increased by the "strategic interruption" of the narrative at crucial moments, while the delay in satisfying the reader's curiosity was prolonged. By structuring the text around the gaps between installments and by making those gaps literally days in length, the serial novel supercharged the reader's imagination and made him or her a more active reader.[8]

The relationship Iser sees between "strategic interruption" and heightened enjoyment would seem to apply with particular force to the experience of watching soap operas. It might also be responsible, in part at least, for the frequently commented-upon loyalty of many soap opera viewers and for the pleasure many viewers take in talking about their "stories" (my mother's generic term for soap operas) with other viewers. The day-long, institutionally enforced suspension of those stories increases the viewer's desire to once again join the lives of the characters the viewer has come to know over the course of years of viewing. And, because the viewer cannot induce the text to start up again, some of the energy generated by this protensive tension might get channeled into discourse about the text among fellow viewers. Furthermore, the range of protensive possibilities the viewer has to talk and wonder about is considerably wider in soap operas than in many other types of narrative. Unlike texts with a single protagonist with whom the reader identifies almost exclusively, the soap opera distributes interest among an entire community of characters, thus making any one character narratively dispensable. Even characters the viewer has known for decades may suddenly die in plane crashes, lapse into comas, or simply "leave town."

Textual gaps exist not only between soap opera episodes but within each episode as well. Each episode is planned around the placement of commercial messages, so that the scene immediately preceding a commercial raises a narrative question. For the sponsor, the soap opera narrative text is but a pretext for the commercial—the "bait" that arouses the viewer's interest and prepares him or her for the delivery of the sales pitch. For the viewer, however, the commercial is an interruption of the narrative—another gap between textual segments, providing an excellent opportunity to reassess previous textual information and reformulate expectations regarding future developments. We might even argue that the repetition and predictability of commercial messages encourages this retentive and protensive activity. The Tide commercial might be novel enough to attend to the first time, but is not likely to hold the viewer's attention thereafter.

Iser theorizes that textual gaps can also be created by "cutting" between plot lines in a story. Just when the reader's interest has been secured by the characters and situation of one plot line, the text shifts perspective suddenly to another set of characters and another plot strand. In doing so, says Iser, "the reader is forced to try to find connections between the hitherto familiar story and the new, unforeseeable situations. He is faced with a whole network of possibilities, and thus begins himself to formulate missing links."[9] As regular soap opera viewers know, in any given episode there are likely to be three, four, or more major plot lines unfolding. The text "cuts" among them constantly. The action in scene 1 might simply be suspended for a time while we look in on another plot line. Later in the episode we might rejoin the action in scene 1 as if no time had elapsed in the interval, or we might join that plot line at a later moment in time.

The gaps that structure the soap opera viewing experience—between episodes, between one scene and the next, as well as those created by commercial interruptions—become all the more important when one considers the complex network of character relationships formed by the soap opera community. In a sense, the soap opera trades narrative closure for paradigmatic complexity. Anything might happen to an individual character, but, in the long run, it will not affect the community of characters as a whole. By the same token, everything that happens to an individual character affects other characters to whom he or she is related.

When I first began watching soap operas regularly I was struck by the amount of narrative redundancy within each episode. In scene 1 Skip tells Carol that he is calling off the wedding. Two scenes later,

Carol tells Greg that Skip has called off the wedding. After the first commercial break, Greg tells Susan that Skip and Carol have broken up. This same piece of information—that Skip and Carol are not to be wed—might be repeated four or five times in the course of an hour-long episode. The repetition of information from one episode to the next can be accounted for as an attempt to keep infrequent viewers up to date, but why is the same piece of information related many times within the same episode? This is a puzzle only for the new soap opera viewer. The regular viewer, familiar with the paradigmatic structure of that particular soap (that is, its network of character relationships), will know that *who* tells *whom* is just as important as *what* is being told. The regular viewer knows that Greg still loves Carol and that Susan has schemed to keep Skip and Carol apart so that she can have Skip for herself. Each retelling of the information, "Skip has called off the wedding," is viewed against the background formed by the totality of character interrelationships. Thus the second and third retellings within the same episode are far from being *paradigmatically* redundant.

How is this paradigmatic complexity related to the structuring gaps of the soap opera text? The size of the soap opera community, the complexity of its character relationships, and the fact that soap opera characters possess both histories and memories all combine to create an almost infinite set of potential connections between one plot event and another. The syntagmatic juxtaposition of two plot lines (a scene from one following or preceding a scene from the other) arouses in the viewer the possibility of a paradigmatic connection between them. But because the connection the text makes is only a syntagmatic one, the viewer is left to imagine what sort of, if any, other connection they might have. The range of latent relationships evoked by the gaps between scenes is dependent upon the viewer's familiarity with the current community of characters and his or her historical knowledge of previous character relationships. In a very real sense, then, the better one "knows" a soap opera, the greater reason one has for wanting to watch every day. Conversely, the less involved one is in a given soap opera's textual network, the more that soap opera appears to be merely a series of plot lines that unfold so slowly that virtually nothing "happens" in any given episode and the more tiresomely redundant each episode seems.

HOW DOES THE TEXT ATTEMPT TO CONTROL THE READING ACT?

In 1961 Wayne C. Booth's book *The Rhetoric of Fiction* foregrounded a common-sensical but frequently overlooked fact of literature: every story represents not just the construction of a fictional world, but a story told from a certain perspective by certain narrational means. In other words, as Sarah Kozloff points out in her essay, every story implies a storyteller. Furthermore, the perspective from which the story is told carries with it attitudes toward the fictional world created in the story. Even in works that ostensibly give us "just the facts," inevitably there are attitudes toward those facts given as well.

Reader-oriented criticism has focused on the corollary to Booth's observation: if every story necessarily involves a storyteller, it also involves someone to whom and for whose benefit the story is being told. But before proceeding further, we need to establish the nature of the storyteller and story reader we're talking about here. Obviously, to say that a piece of narrative fiction involves a communication between addresser and addressee is to compare it to a face-to-face communication. However, reading a novel or watching a television drama differs in several key respects from talking with a friend. In the first place, the actual addresser (author) of a piece of narrative fiction is absent at the moment of reading, just as the actual addressee (reader) is always absent at the moment of writing. Furthermore, what is being referred to in a novel or fictional television drama is only indirectly related to the immediate environment either the author or the reader inhabits (it is the fact that what we are reading does not exist that makes it fiction), so that the reader has no "real" points of reference against which to test the message being delivered in the narrative. Thus both the addresser and the addressee dealt with by narratological analysis and (with several notable exceptions to be discussed later) reader-oriented criticism are textual constructs, not flesh-and-blood human beings. To use a dangerously ambiguous term, the reader of a story constructs its teller as the "point of view" with which the values, norms, and attitudes expressed in the work are consistent.

In attempting to specify "to whom" a story is told and the role this addressee ought/might play in the reading process, reader-oriented theorists have proposed a bewildering array of readers: "fictive reader"

(Iser), "model reader" (Eco), "intended reader" (Wolff), "characterized reader" (Prince), "ideal reader" (Culler and others), "inside" and "outside reader" (Sherbo), "implied reader" (Iser, Booth, Chatman, and others), and "superreader" (Riffaterre)—to name but some of the "readers" who now populate literary studies. Although each of these terms constructs a reader different in some respects from all the others, all of them refer to one of two types of readers: "implied fictional readers" and "characterized fictional readers." (Yes, I know I just added two more—and more complicated—terms to the list, but you can't very well talk about "two types of readers" without naming them in some fashion.)[10]

As anyone who has ever tried to tell an anecdote or a joke knows, every story is constructed around a set of assumptions the teller makes about his or her audience: what they know or don't know; how they are likely to feel about certain things (Republicans, college teachers, mothers-in-law); why they are willing to listen to the story to begin with; how it is likely to fit in with other stories or jokes they might already have heard; and so forth. "Model," "ideal," "super," "implied," and "intended" all refer to the composite of these assumptions as they are manifested within the narrative itself—hence the term "implied fictional reader," which I am borrowing from W. Daniel Wilson to refer to this category of reader.

For some theorists, the implied fictional reader is a projection into the text of the qualities possessed by the kind of reader the author had in mind when he or she wrote the work—hence the use of the terms "model," "ideal," and "super" to refer to this reader in the author's mind. In semiotic terms, the ideal reader would be one who fully shared the textual, lexical, cultural, and ideological codes employed by the author. As Umberto Eco describes the model reader, it is the reader "supposedly able to deal interpretively with the [text's] expressions in the same way as the author deals generatively with them." Eco's formulation reminds us that for any narrative to "work" for the reader, he or she must literally and figuratively speak something resembling the same language as the author.[11] The author must be able to presuppose certain competencies on the part of the reader if for no other reason than in order to select those things that must be made explicit in the text from those that can be left unsaid, knowing that the reader should be able to fill them in.

Some reader-oriented critics would make the correlation between the ideal reader and real readers a touchstone for intepretive validity. They would argue that the closer the reader comes to the qualities possessed by the ideal reader the author had in mind when he or she wrote the work, the closer that reader comes to fully understanding the work. As Booth puts it, "The author creates . . . an image of himself and another image of his reader; he makes the reader, as he makes his second self, and the most successful reading is one in which the created selves, author and reader, can find complete agreement."[12] Others, however, would claim that although the author might have kept before him or her the image of an ideal reader, the only real reader that could possibly fill that role is the author him- or herself! Were the author and reader to share *every* code, literary communication would be unnecessary because there would be nothing new to say.

In an attempt to avoid getting too bogged down in the ongoing debates over the nature of and name for the implied reader in the text, we might regard the implied fictional reader as a textual place or site rather than a hypothetical person. It is the position the text asks us to occupy—the preferred vantage point from which to observe the world of the text. In other words, every fiction offers not only a structure of characters, events, and settings, but a structure of attitudes, norms, and values as well. The reader is invited to take up a position relative to these structures. That "place" turns out to be the other side of the point of view that has organized the text's world—the point of view the reader is offered from which to observe that world.

One of the most obvious ways the reader's place in the text can be established is by referring directly to the reader: addressing the reader directly, confiding in the reader, appealing to the reader, describing what the reader knows or probably feels, even questioning or challenging the reader's interpretation of the text thus far. In other words, the text might create a "characterized fictional reader." Such a strategy was common in the eighteenth-century British novel (Fielding's *Tom Jones,* for example), reaching perhaps its most elaborate (and funniest) use in Laurence Sterne's novel, *Tristram Shandy.*[13]

The characterized reader was less frequently employed in nineteenth-century novels, and by the twentieth century had all but disappeared from mainstream fiction. Similarly, by the time we reach Hemingway, narrators in novels had become, if not invisible, certainly depersonalized.

The classical Hollywood cinema expends tremendous effort to hide the means by which it tells its stories. It also engages its viewers covertly, making them unseen observers of the world that always appears fully formed and autonomous. With very few exceptions (most of them comedies), the viewer of a Hollywood film is neither addressed or acknowledged. One of the cardinal sins of Hollywood acting style is looking into the lens of the camera, because doing so threatens to break the illusion of "as if it were real" by reminding viewers of the apparatus that intervenes between them and the world on the screen. This is certainly not to say that there is no implied viewer constructed by Hollywood films. In fact, if we once again consider the implied viewer as a position relative to the world of the text, it is easier to see how the viewer is located "within" the Hollywood film than is the reader within the traditional novel. Given that the viewer's knowledge of the world of the film comes through the camera, the viewer is quite literally positioned "some place" relative to the action in every shot. Furthermore, as the chapters on feminist and psychoanalytic criticism make clear, the Hollywood film carefully "writes" its viewer into the text, usually establishing its gender as male and skillfully regulating the manner by which "he" engages with the characters.

Although we would find being addressed directly by a character in a Hollywood film unusual (and perhaps discomforting), as Sarah Kozloff points out we accept the "characterized viewer" as an integral part of television. In fact, our experience of television involves two quite different modes of viewer engagement. The first, which we might call the Hollywood narrative mode, represents the adaptation to television of the classical Hollywood narrative style and its means of drawing the viewer into the text: spectator omniscience and invisibility, alternation between third- and first-person points of view and between shot and reverse shot, hiding the means by which the world of the text is created, etcetera. We find the Hollywood mode most prevalent in those television programs still shot "film style"—prime-time dramas of all sorts, (including both series and serials), made-for-TV movies, and some situation comedies. Daytime soap operas and some situation comedies have modified the Hollywood style to accommodate what is called "three-camera, live-tape" shooting; an entire scene is enacted while being shot simultaneously by three (or more) television cameras. The director electronically "cuts" between one camera and another as the scene unfolds. Live-tape production makes the shot/reverse shot and subjective point-of-view shot much more difficult to achieve than in Hollywood-style filmmaking, because repositioning the camera for the reverse shot would require penetrating the space of the scene. Hence, subjectivity in soap operas is usually rendered aurally rather than visually, by showing a close-up of a character while his or her thoughts are heard on the sound track. Despite some degree of deviation from the Hollywood cinema style, however, live-tape television style seldom, if ever, addresses the viewer and observes most other conventions of Hollywood style.

The other mode of viewer engagement on television, which we might call the rhetorical mode, derives historically from radio practice of the 1930s and 1940s and ultimately, I suppose, from the "point-to-point" communication technologies that radio superceded: the telephone and the telegraph.[14] Included in the rhetorical mode would be news programs, variety shows, talk shows, "self-help" and educational programs (cooking, exercise, and gardening shows, for example), sports, game shows, and many commercials. In the rhetorical mode, both the addresser and the addressee (what Sarah Kozloff calls the "narratee") are openly acknowledged. The former is frequently personified or "characterized" as the reporter, anchorperson, announcer, host, master of ceremonies, or quiz master. The viewer is addressed directly as characters look directly into the camera and speak to "you, the home viewer." The means of presentation, particularly the technological means of presentation, are frequently emphasized rather than hidden. David Letterman shoots rubber darts at the camera; *Jeopardy* "answers" are revealed via a bank of television monitors; local television newscasts originate from what looks to be the newsroom itself; television screens are built into the sets of news programs so that anchorpersons can talk with reporters in the field; and (in our local television market, at least) a favorite closing shot for the evening newscast is one that reveals set, cameras, and all. The "personified addresser"—host, anchor, or quiz master—manipulates this technology and mediates between the world on the other side of the screen and that in our living rooms. In doing so, he or she (it is usually a he) offers "us" a better view, more information, a dazzling technical spectacle, or, seemingly, an insider's view of how things "really work" behind the scenes.

One of the hallmarks of the rhetorical mode—and another striking difference between its method of viewer engagement and that offered by Hollywood films—is its use of characterized viewers. Direct address is but the most obvious way by which the viewer is represented on television (as the "person" Dan Rather says "good evening" to at the beginning of *CBS Evening News*). Television frequently provides us with on-screen characterized viewers—textual surrogates who "do" what real viewers cannot: interact with other characters and respond in an ideal fashion to the appeals, demands, and urgings of the addresser.

These on-screen characterized viewers abound on television commercials. An ad for *Time* magazine, for example, opens with a man sitting at his desk at home. An off-screen voice asks him, "How would you like to get *Time* delivered to your home every week for half-off the newsstand price?" The man looks into the camera as the voice speaks, but before he can respond the voice adds, "You'll also receive this pocket calculator with your paid subscription." An arm emerges from off-screen and hands the calculator to the man. He nods his acceptance of the offer, but before he can speak the voice piles on still more incentives. Finally, with not the slightest doubt remaining that the man will become a *Time* subscriber, the voice orders him to place the toll-free call. The man hesitates. "What are you waiting for?" the voice asks. "You haven't told me the number," the man objects. The voice responds with the number and it magically appears at the bottom of the screen. The commercial ends with the man placing the phone call.

Notice that in this example the characterized addressee stands in a different relationship to the text's addresser than does the implied addressee (the "presumed" viewer at home). The man in the *Time* ad enjoys a direct, face-to-face (or, in this case at least, face-to-voice) relationship with the person who addresses him. The technology necessary to bring the commercial message "to us" disappears and is replaced by an unmediated interpersonal communication situation. In many television commercials and network and local promos, the characterized addressee is established not just to personalize and textualize the implied viewer but to make an interpersonal exchange out of a one-way, mass-communication phenomenon. The characterized addressee is established in a setting suggesting that of the implied audience: the kitchen (particularly in commercials directed at women and shown during daytime programming), the den, the family or living room. Then the addresser enters the space of the characterized addressee and talks with him or her directly. In a Drano ad, for example, we see an aproned woman standing forlornly in front of her sink while a male voice booms accusingly, "Y O U R S I N K' S S T O P P E D U P." A few years ago, at the beginning of the fall season, CBS ran a series of promotions featuring some of its schedule's most famous stars. In one vignette, a weary young woman returns to her apartment after a hard day at work, turns on the television set, and is startled to find Tom Selleck, as Magnum, standing in her living room. In a parallel scene, a young man turns on his TV set, and a television image of Loni Anderson magically is transformed into the actress herself in the man's den.

The *Time* ad illustrates another aspect of television's use of the characterized viewer—a blurring of the distinction between characterized addressee, implied addressee, and addresser. When the man responds to the voice, he does so by looking directly into the camera. Thus, he looks at "us" as if we were the source of the message. In a Hollywood film one of the principal ways of establishing identification between the viewer and a character in the film is via a strategy called "glance/object" editing. We are shown a close-up of a character as that character looks off-screen. The second shot, taken from that character's point of view, shows us what he or she sees. A third shot returns us to the close-up of the character. In the rhetorical television mode, however, glance/object editing is short-circuited, because "we" turn out to be the object of the character's glance. In the curious logic of this mode, the "voice" of the commercial is made into our voice, as the man establishes the connection between our gaze and "the" voice. At the same time "we" are characterized as the man who responds to that voice, he who acts as we "should" act. The superimposition of the telephone number at the end of the ad, however, addresses "us" rather than "him," because 1) he attends to the voice telling him the number, rather than to its appearance on the screen, and 2) even if he did notice it, it would be backwards!

This purposive collapsing of addresser, characterized addressee, and implied addressee in television's rhetorical mode creates what Robert Stam has called, with regard to news programming, "the regime of the fictive We." In the middle of *As the World Turns*, the announcer says, "We'll return in just a moment." A promo for *PM Magazine* says "Tonight on *PM Magazine* we'll journey to Rome. . . ." The examples are legion. Who is this "we"? Perhaps it merely stands for the collective "senders" of the message—the news staff, the "folks" at Procter and Gamble, and so on. But the referent of television's "we" is usually left vague enough to cover both the addresser and the implied addressee. Stam sees the "misrecognition of mirror-like images" in the fictive We to have serious political consequences:

> Shortly after the ill-fated "rescue attempt" in Iran . . . Chuck Scarborough of New York's *Channel 4 News* began his newscast, "Well, we did our best; but we didn't make it." The "We" in this case presumably included the newscaster, the president, and a few aides. It certainly did not include the majority of Americans, even if their "support" could be artfully simulated after the fact. Television news, then, claims to speak for us, and often does, but just as often it deprives us of the right to speak by deluding us into thinking that its discourse is our own.[15]

The characterized addressee plays an equally important role in two other television genres—the game show and the talk show. Whereas the commercial and the news program tend to characterize their addressees individually, game and talk shows represent their addressees as a group—the "studio audience." Also, to follow up a point made by Kozloff, in the examples above the impersonal experience of watching television was made into an interpersonal exchange by situating the action on the "viewer's" side of the television set, in the characterized addressee's living room, kitchen, etcetera. In talk and game shows, the characterized viewer is made a part of "the show." The studio audience is "there," where it really happens, able to experience the show with their own eyes rather than through the mediation of the camera lens.

Most game and talk shows carefully regulate the response of their studio audiences in order that this "actual" audience is characterized to the home viewer as an ideal audience. With the aid of "applause" signs in the studio, the audience unfailingly responds at the appropriate moment—when a new guest is introduced, when the contestant wins the big prize, when it is time for a commercial. Some game and talk shows employ someone (who usually doubles as the show's announcer) to lead the studio audience's response. Occasionally on *Late Night with David Letterman*, for example, the viewer at home can see announcer Bill Wendell standing between studio audience and set frantically waving his arms for a more vociferous response from the studio audience. As with many talk shows, Wendell also "warms up" the studio audience before taping begins by telling a few jokes, informing the audience what to expect when taping actually begins, and quite literally telling them how to behave as a "good" audience. At a taping I attended a few years ago, Wendell admonished us, "If you chuckle silently to yourself, no one at home will think you're laughing at all. When you think something is funny, let it out!"

Game and talk shows also employ devices to individualize the studio audience. David Letterman and Johnny Carson go into the audience to play "Stump the Band" or "Ask Mr. Melman" with selected members of the audience. *Donahue* is predicated upon individual members of the studio audience asking "guests" the type of questions "we" would ask if "we" were there. Notice, however, that even when the characterized viewer is allowed to speak as an individual member of the studio audience, his or her discourse is carefully regulated and channeled. It is Phil Donahue (or his assistant) who wields the microphone and determines who is chosen to speak and for how long. The person speaking speaks to and looks at either Phil or the guests on stage. Only Phil looks directly into the camera and addresses "us." Notice also that by positioning Phil among the studio audience rather than on stage with the guests and by allowing him to serve as a spokesman for both studio and home viewers ("You'll forgive us, Mr. X, if we are just a little skeptical of your claim that all we need to do to balance the budget is . . ."), *Donahue* collapses "host," studio audience, and home audience into television's fictive "we," and covers over the means by which the responses of the characterized viewer are regulated.

In the talk show, although the studio audience is addressed and individual members are allowed to speak, the roles of host, guests, and characterized audience are demarcated, if on some shows purposefully blurred. The audience stays "in its place" on this side of the stage; guests are isolated onstage in front of the audience (at home and in the studio); the host negotiates and regulates the relationship between them and the home viewer. Except in the unlikely event that a studio audience member is called upon to speak for a few seconds (even on *Donahue* only a tiny portion of the studio audience actually speaks), his or her role is primarily that of exemplary viewer—one who listens, looks, and responds appropriately. In the game show, however, the characterized viewer crosses the line normally separating characterized "audience" from "show." This transformation of audience member into character is perhaps best exemplified by Johnny Olson's invitation to "come on down" on *The Price Is Right*. We might speculate that a large measure of the pleasure we derive from game shows stems from the fact that the contestant is more like "us" than like "them." As a characterized viewer he or she appears to us as a "real" person acting spontaneously, not an actor reading lines. (Although on many game shows the contestants are carefully screened and coached, and even if drawn at random from the studio audience, the contestant is no doubt aware of the role he or she is expected to play from having watched the show before.) In those instances in which contestants are selected from the studio audience, they are plucked from among "us."

By splitting off one or more characterized viewers from the rest of the studio audience, the game show sets up a circuit of viewer involvement. When Bob Barker asks the contestant to guess how much the travel trailer costs, we almost automatically slip into the role of contestant, guessing along with him or her. If we guess correctly along with the contestant, the bells and whistles go off for us as well as for him or her. But we can also distance ourselves from the contestants and take up the position of the studio audience as they encourage the contestants and, on *The Price Is Right,* at least, shout out what they believe to be the correct guess. As we watch a game show, we constantly shift from one viewer position to another, collapsing the distance between contestants and ourselves as we answer along with them, falling back into the role of studio audience as we assess contestant prowess and luck (or lack thereof), assuming a position superior to both when we know more than they. The viewer-positioning strategy of the game show encourages us to mimic the responses of the characterized viewer in the text. Indeed, I find it difficult to watch a game show *without* vocally responding (whether or not someone else is in the room with me). I can't resist answering the questions myself, nor can I resist commenting on a contestant's abysmal ignorance when I have the correct answer and he or she doesn't.

It is not coincidental that commercial television has developed a sophisticated rhetorical mode of viewer engagement within which so much energy is expended in giving the viewer at home an image of him or herself on the screen. The Hollywood cinema style has developed to serve a system of economic exchange in which the viewer pays "up front" for the opportunity to enjoy the cinematic experience that follows the purchase of a ticket. No further action is required of the viewer once he or she leaves the theatre to fulfill the implicit contract between institution and viewer/consumer. On the other hand, commercial television succeeds only by persuading the viewer to respond at another time, in another place, in a prescribed manner—in other words, the implicit contract the viewer has with television is fulfilled not in front of the television set, but in the grocery store. Television demands that we act; hence it is inherently rhetorical. Furthermore, that action must occur on "this" side of the television set. If the Hollywood film-viewing situation is centripetal (the one bright spot of moving light in a dark room that draws us into another world and holds us there for ninety minutes), then television is centrifugal. Its texts are not only presented for us, but directed out at us. Ironically its messages drive us away from the set, out into the "real" world of commodities and services. By conflating addresser and addressee under the regime of the fictive "we," commercial television softens the bluntness of its rhetorical thrust. By positioning "us" in "their" position, we seem to be talking ourselves into acting. By adopting the style and mode of address of commercials, other genres of television programming rehearse "for fun" what the commercials do in earnest. Every commercial is an implicit unanswered question—"will you buy?"—that calls for action the commercial text itself cannot provide, because only "real" viewers can buy the very real commodities the commercials advertise. By offering characterized viewers within the text, commercials fictively answer their own questions with resounding affirmation. We should not be too surprised, then, when talk shows, game shows, religious programs, and other forms of commercial television programming also "write in" their own viewers and provide them with opportunities to respond and act in an affirming, if carefully regulated, manner.

In this regard, perhaps the ultimate expression of television's rhetorical mode is the television "phone-a-thon," in which appeals for financial contributions to religious groups, public television stations, or charitable causes are made. The host usually stands in front of a raised bank of tables, behind which staff members or volunteers sit answering telephones. The calls they receive are, of course, pledges phoned in from viewers who have responded to the host's plea for financial support. Obviously, it is hardly necessary that the reception of telephone pledges be done on-screen. But fund raisers realize the important function served by keeping those answered telephones in the background. Each ringing telephone is (in semiotic terminology) an indexical sign for a characterized viewer—a reminder to the audience of how they "should" act. Furthermore, by combining televisual and

telephone technologies, the phone-a-thon links the sign of these characterized viewers to the responses of *real* viewers as the latter provide
a material response to the text's rhetorical demand.

TO WHAT DEGREE ARE THE READER AND
THE CONTEXT OF THE READING ACT
DETERMINATIVE OF MEANING?

Reader-oriented approaches to literature would deprive the text of its
power to enforce meaning on the reader and, at the very least, shift
critical attention away from "the words on the page" to the interaction
between reader and textual structure. But does this interaction result
in the reader merely following the text's instructions for the assembly
of meaning? Or is meaning production so dependent upon the individual reader and his or her reading activity that each reading quite literally produces a different meaning? To what degree is meaning production an individual activity at all? How is an individual reading
influenced by forces external to that act—linguistic, cultural, institutional, and ideological forces of which the individual reader might not
even be aware?

Any attempt to grapple with these questions requires that the critic
first answer another: "What, exactly, is the reader-oriented critic analyzing?" For the traditional "figure-in-the-carpet" critic, the object of
study was self-evident. Because the text's meaning was to be found
within it, then obviously critics studied texts. But if we accept, with
Jonathan Culler, that the study of "literature" is no longer an attempt
to interpret texts but to account for "their intelligibility: the ways in
which they make sense, the ways in which readers have made sense of
them," then what are literary critics now supposed to study?[16] Can this
sense-making process be extrapolated from the text itself? As we have
seen in previous sections of this chapter, reader-oriented critics
aligned with the phenomenological school of Ingarden, Iser, and Jauss
would answer a qualified "yes." They would assert that we can learn a
great deal about the reading act through study of the textual structures and strategies that guide and, to some degree, regulate that
process.

Other reader-oriented critics would argue that the critical moves of
Iser and others represent a feint rather than a shift in the orientation of

literary studies. Given that Iser's "reader" is always an abstract construct, evidence for whose alleged maneuvers are always drawn from something we are supposed to recognize as "the text," how different, they ask, is Iser's analysis of the reading act from more traditional types of interpretation?[17] Has he merely replaced "the" meaning of a text with a slightly more liberal notion of "instructions for meaning production?"

Terry Eagleton, among others, points out the paradox in Iser's theory caused by his acceptance of the proposition that meaning is a product of a reader's activity at the same time he holds to the idea that evidence for that activity can be found by studying texts. "If one considers the 'text in itself' as a kind of skeleton, a set of 'schemata' waiting to be concretized in various ways by various readers, how can one discuss these schemata at all without having already concretized them? In speaking of the 'text itself,' measuring it as a norm against particular interpretations of it, is one ever dealing with anything more than one's own concretization?" Eagleton calls this a version "of the old problem of how one can know the light in the refrigerator is off when the door is closed."[18]

Iser's counter here has been to maintain a certain degree of objectivity for the textual features that guide the reader's response. Without this, he argues, the text melts away into the subjective experience of each reader's activation of it. In other words, Iser worries that if we can't agree there's something "there" in the text that at least stimulates and guides meaning production, literary scholars might as well pack up shop and find another line of work.

For several American reader-oriented critics, Iser's paradox is created by his unwillingness to accept the consequences of his own line of reasoning—even if taking that step means that literary scholars have to find something to study other than "the text." For Norman Holland, the reading act is a kind of transactional therapy and process of individual self-realization. In reading, he says, the reader uses the text "as grist with which to re-create himself, that is, to make yet another variation of his single, enduring identity."[19] This "single, enduring identity" constitutes each personality's central organizing principle and the means by which he or she relates to the outside world. This identity changes as the individual adjusts to the flux of experience, but it also structures and reorders that flux according to its internal psychic logic. Holland uses the term "work" of literature as a verb rather than a noun. The words on the page merely initiate a process by which the reader "works on" a story, "works it over" according to his identity theme, and, finally, "works it into" his or her identity.

If we accept Holland's account of how texts (now reduced to merely words on the page) are made to mean, then the object of study for the literary critic becomes the reader's identity theme and the manner by which it "works" on texts, because it—not the text—is the stable and determining feature of the reading transaction. In 5 *Readers Reading,* Holland puts his theory into practice. He identifies the identity themes of five undergraduate students through standard personality tests and other means, has them all read several stories, and then discusses their responses with them to attempt to discover signs of their individual identity themes in their interpretations of the stories. What results is, for Holland, proof of his theory of reading as self-realization. For Holland's detractors, however, 5 *Readers Reading* is either a circular and self-fulfilling exercise (by establishing the "identity theme" as the independent variable in his experiment and then looking for evidence of its effect in the reader's responses, Holland is almost sure to find what he is looking for) or an inadvertent demonstration that, if Holland's theory *is* correct, the critic can do no more than reflect upon his or her own identity theme, not anyone else's. As William Ray puts it, "the subject whose self re-creation fills 5 *Readers Reading* is none of the five students described, but Holland himself, whose hundreds of pages of interpretation have predigested and reformulated the raw data of the reader's responses and therefore represent the only sustained reader transaction to which we could possibly respond."[20]

For Stanley Fish, reading is not the individual and self-realizing activity Holland describes, but a fundamentally social process. In his view, the "self" is "a social construct whose operations are delimited by the systems of intelligibility that inform it."[21] Thus differences in interpretation arise from differences in the assumptions that underlie different "interpretive communities," rather than from differences between individuals. What appears to the reader as his or her individual imposition of meaning is actually the result of a system of belief and resultant interpretive strategies he or she shares (usually unknowingly) with a larger community of readers. Fish shifts the focus of literary criticism one step further away from "the text." According to him, there are no "textual structures" that exist apart from a particular interpretive strategy that looks for and values them; there is no individual reader whose activities might be isolated and theorized; both text and individual reader are social products. With Fish the very notion of a text depends upon an interpretive community that endows this set of marks with that status. The famous example Fish uses in his *Is There a Text in This Class?* is of a reading list left on the blackboard of his classroom by a previous instructor. Told by Fish that this list was a poem, the class was able to produce analyses of it *as* a poem—thus demonstrating to Fish that the category "poetry" is a function, not of authorial intention or textual organization, but rather of a particular system of belief (in this case that which governs reading in college literature classes).[22]

What, then, does Fish suggest literary critics study? Implicit in Fish's formulation of reading is the challenge to the critic to uncover the largely unarticulated assumptions of interpretive communities, especially those that dominate the reception of texts at a particular point in history, which would presumably account for variations in reading between readers and across cultures and time. Fish is quick to point out, however, that such an exercise is also subject to the forces it attempts to explain. Unlike Holland, who would reserve some objective space within which the critic could operate, Fish fully admits that his ability to read other critics as manifestations of the beliefs of a particular interpretive community is only possible because of the beliefs that inform Fish's own work—beliefs that, in effect, sanction what Fish does as "criticism." As with other reader-oriented theories, one has to admit that there is no position "outside" reading from which one can read the responses of other readers. Fish holds out no hope to those who would adopt his metacritical strategy that the uncovering of the assumptions of another interpretive community (or even of the critic's own) will change anyone's mind or bring us closer to "the truth." He warns his readers that his theory "is not one that you (or anyone else) could live by. Its thesis is that whatever seems to you to be obvious and inescapable is only so within some institutional or conventional structure, and that means that you can never operate outside some such structure, even if you are persuaded by the thesis."[23]

Fish's emphasis on the power of interpretive communities to determine the meaning of texts (and, indeed, the power to establish what a text is) should serve as one more reminder that all reading activity, including "reading" television, occurs within larger contexts. Furthermore, although he differs with them in many other respects, Fish shares with semiotic, ideological, and pyschoanalytic theorists the recognition that language, ideology, culture, and institutions to some degree "speak" us as readers and viewers. One of the "axioms" of modern critical research, says Jonathan Culler, is "that the individuality of the individual cannot function as a principle of explanation, for it is itself a complex cultural construct, a heterogeneous product rather than a unified cause."[24]

As students of television, we might also be reminded by Fish's notion of "interpretive communities" of the norms of the critical community by which television has been judged as an actual or potential art form. As I pointed out in the Introduction, one of the reasons, I believe, that so little aesthetic analysis of television has been produced over the past forty years (relative to the pervasiveness of the medium and the sheer quantity of "texts" produced over that same time) is that the dominant critical community has embraced the values of what Iser would call "traditional" criticism. In addition to the "figure-in-the-carpet" assumptions about what and how texts mean, these values would include the belief that a single artistic vision should be expressed in a work of art, that good art requires intellectual work on the part of the perceiver, and that art works should be autonomous and unified entities, among others. As I have argued elsewhere with specific reference to radio and television soap operas, it is difficult to fit many forms of broadcast programming into a critical schema that assumes these values.[25] Who is the author of a situation comedy? How can one analyze a text that, like the soap opera, refuses to end? Because critics have found it difficult to discern the textual markers of art in television, they have seen little there to analyze. Furthermore, where critics have found the exceptional program on television worthy of detailed analysis (usually a dramatic program that shares some of the qualities of literary or theatrical "art"), they have spoken of it as if its sterling qualities were inside the work itself and "found" by the perceptive critic. This strategy makes it appear that the standards the critic applies are universal and renders evaluative assessments properties of the work, not the application of critical norms by the critic. As Fish points out, however, "all aesthetics . . . are local and conventional rather than universal, reflecting a collective decision as to what will count as literature. . . . Thus criteria of evaluation (that is, criteria for identifying literature) are valid only for the aesthetic they support and reflect."[26]

Another reader-oriented critic who foregrounds the social nature of the reading act is Tony Bennett. Following on the work of Pierre Macherey, Bennett has contended that all texts come to the reader always already "encrusted" with the effects of previous readings and that it is pointless to speak of a "text" existing separate from these historically specific encrustations. Our reading of any work is inevitably conditioned by other discourses that circulate around it: advertisements for it, reviews, other works of the same genre or author, etc. For Bennett, studying the relationship between literary phenomena and society "requires that everything that has been said or written about a text, every context in which it has been inscribed by the uses to which it has been put, should, in principle, be regarded as relevant to and assigned methodological parity within such a study."[27]

Bennett cites the example of the James Bond phenomenon. Our viewing of a given James Bond film is conditioned by the previous films in the series; by the novels upon which the films are based; by the characterizations of Bond by Sean Connery and Roger Moore; by the advertisements for both novels and films; by the cover designs for the novels (featuring scantily clad women and the paraphernalia of espionage); by the songs written to accompany the films; by articles in the press about the films, their stars, Ian Fleming, "Bond" himself, British intelligence, and so forth. The film itself is merely one part of "a mobile system of circulating signifiers," a text that is activated by the reader only in relation to the reading of other texts.[28]

Commercial television might be seen as a gigantic "mobile system of circulating signifiers." Perhaps more than any other form of cultural production, television produces texts that never "stand alone." Rather, they continuously point the viewer in the direction of other texts. This is not surprising, considering the economic basis of commercial television. The viewer does not "buy" the right to view a program (as the reader buys a novel or a moviegoer buys a ticket to a film). The viewer is, in fact, sold to advertisers in lots of one thousand. The text's (program's) status in this exchange is merely that of the bait to keep the viewer watching until the commercial comes on. To the commercial networks, each program is but one device in a larger scheme designed to hold the viewer's attention throughout a block of the day.

Among television program types, the made-for-TV movie might appear to be among the more autonomous with respect to its relationship to other texts.[29] It is narratively self-contained, and, unlike the series or the serial, there are no characters or situations that carry over from one movie to the next. In fact, however, the made-for-TV movie relies heavily upon the viewer's familiarity with other texts.

Since the mid-1960s, television programmers have used the made-for-TV movie as an economical substitute for theatrical feature films. The made-for-TV movie presents a promotional problem not found with either the theatrical feature film or the television series, however. Having played in theatres a year or so before its television air date, the theatrical feature film is very much the "already read" text, bringing with it a trail of discourse that accompanied its theatrical release. Thus the audience's familiarity with the film's title, theatrical success, stars, etcetera, make the theatrical feature a known quantity by the time it reaches television screens. Carrying over its characters and setting from week to week, the series has a built-in self-promotional aspect: each episode acts as a "preview of coming attractions" for the next episode. Having no prior theatrical "life" and limited to a single airing, each made-for-TV movie presents a promotional challenge. How can an audience be built for a film whose budget usually precludes the use of big-name Hollywood stars or being based upon a popular novel, and whose single exposure on television means there can be no "word of mouth"? What is there to promote and advertise?

A number of strategies have been employed in an attempt to solve this problem, one of the most common being to make the made-for-TV movie a "problem picture." The film is based upon a current social controversy, and discourse about that controversy is used to promote the film. Over the past ten years nearly every social issue imaginable has become the basis for a made-for-TV movie: missing children (*Adam*), spouse abuse (*The Burning Bed*), incest (*Something about Amelia*), teenage suicide (*Surviving*), and child molestation (*Kids Don't Tell*), among many others. The film then represents a narrativization of discourse about a particular social issue already circulating in newspapers, magazines, and other television programs. As Laurie Schulze and others have pointed out, the more disturbing or threatening aspects of these social problems are defused by the made-for-TV movie's reliance upon familiar genres as vehicles for their expression.[30] Typically, whether the problem is teenage prostitution or single bars, it is framed by the conventions of the family melodrama and/or the heterosexual romance. In the made-for-TV movie Schulze analyzes—*Getting Physical,* a 1984 film about female bodybuilding—this generic framing occurs even before the film is aired. One of the most important promotional devices for the made-for-TV movie is display advertising in *TV Guide*. The viewer's interest must be aroused by an ad illustrating the film's "problem" visually (usually by artwork, rather than a still from the film itself) and describing it in a few words of accompanying text. The *TV Guide* ad for *Getting Physical* showed a bikini-clad woman in a frontal bodybuilder pose, beside a photographic insert of a close-up shot of another woman and a man. The text read, "When a beautiful woman becomes a bodybuilder, the sport takes on a whole new shape, and her life new meaning." Schulze comments: "The smaller inserted photograph suggests that the benefit will have something to do with heterosexual desirability. The insert implies that the 'new meaning' given her life will involve the romantic attentions of an attractive male; bodybuilding will facilitate heterosexual romance."[31]

Made-for-TV movies are also promoted on other television programs. In fact, some television talk shows have become little more than vehicles for the promotion of other texts—books, theatrical movies, records, concert appearances, and other television programs. The stars of a forthcoming made-for-TV movie are made "available" for interviews on *Entertainment Tonight, Show Business This Week,* network morning news shows, and other such programs. The interviews may or may not center on the star's role in the made-for-TV movie, but a "plug" for it is sure to be included. Such star appearances are, of course, most likely to occur on the network showing the made-for-TV movie, and the interview's content and timing are carefully orchestrated to give maximum promotion to the movie.

In January 1986, NBC ran a made-for-TV movie entitled *Mafia Princess,* based on the memoirs of the daughter of reputed Mafia leader Sam Giancana. The film starred Tony Curtis and *All My Children* star Susan Lucci. Curtis appeared on both the *Today* show and *Late Night with David Letterman* the week before the film's airing. Curtis's remarks about the film on both programs stressed that it was "really" about a relationship between a father and a daughter, thus helping to frame it within the family melodrama genre. Interestingly, however, the interviews touched relatively briefly on *Mafia Princess,* emphasizing instead Curtis's recovery from drug abuse, his early struggles as a Hollywood actor, and his relationship with his family (particularly with his daughter, Jamie Leigh Curtis). Despite the fact that these topics did not concern *Mafia Princess* directly, they provided additional layers of discursive "encrustation." Reminded of Curtis's battle with drug abuse, the viewer might watch *Mafia Princess* for signs of the actor's rehabilitation. If the viewer had not made the connection between the film's subject matter and Curtis's sometimes stormy relationship with his own daughter before the interviews, he or she could not help but be aware of it afterwards. David Letterman even asked Curtis if he had encountered any "Mafia types" during his career in show business. He responded that he had, but coyly refused to discuss the matter further.

All of these encrustations are in addition to the most obvious—representations of organized crime the audience might have encountered in newspapers, magazine articles, other television programs, and films. Even if a viewer of *Mafia Princess* had managed to miss the ad and article in *TV Guide,* interviews with Tony Curtis and Susan Lucci, and other forms of promotion, he or she still would not have come to it as a "naive" reader. *Mafia Princess* cannot be separated from the web of other texts for which it provides a site of intersection—texts about organized crime, Italians, law and criminality, the historical personages upon whom the story was based, family relationships, and so on. Even this rather unremarkable example of textual encrustation serves to support Bennett's contention that "the text is never available for analysis except in the context of its activations."[32]

CONCLUSIONS

One major difficulty in discussing how reader-oriented criticism might relate to television analysis is the considerable theoretical and methodological diversity among critics whose work might bear the label "reader-oriented." As the section of this chapter on characterized and implied readers and viewers illustrates, there are about as many notions of what a reader is as there are critics talking about readers. Still, the strand of contemporary literary criticism that, however loosely, we can call reader-oriented has helped to at least raise a set of questions that traditional literary analysis left unasked, and in doing so has challenged us to reconsider concepts and assumptions that lie at the very heart of the critical endeavor. What is a text? How is it made to mean? What is the relationship between the world in the text and the world brought to the reading experience by the reader? To what degree is the sense-making capacity of the reader a product of external forces? In a world without texts determinative of their own meaning, what is the role of the critic? Given the fundamental nature of these questions, it should come as no surprise that reader-oriented critics fundamentally disagree about their answers.

The relationship between television and its viewers provides an excellent laboratory in which to test the insights of reader-oriented literary critics—even if, as in the case of Iser, some of those critics themselves might question the applicability of literary theory to the realm of nonliterary popular culture. The movement in reader-oriented criticism away from the notion of a stable and eternal text to that of activations of texts within historically specific conditions of reception is accelerated by the very nature of television. Television programming is inherently ephemeral—here for an hour or so and then gone, perhaps forever. Furthermore, few people in the television industry think in terms of programming as a series of autonomous and isolated texts. Because the goal of commercial television is the stimulation of habitual viewing over long periods, programs are conceived of as links in a continuous chain of programming. Raymond Williams has spoken of television programming as a "planned flow, in which the true series is not the published sequence of programme items but this sequence transformed by the inclusion of another kind of sequence, so that these sequences together compose the real flow, the real 'broadcasting.'"[33]

The insight that texts carry within them a place marked out for the hypothetical reader to occupy applies with particular force to television. Because of its economic nature (that of a vehicle for selling people to advertisers), commercial television addresses its prospective viewers much more directly than does the fictional cinema or literature. The need of advertisers to persuade viewers to become "good viewers" (to accept the arguments of the commercial messages and then purchase the product) infuses all aspects of this "flow" of programming. Viewers are not only directly addressed, they are provided with representations of themselves on the screen and embraced within the all-encompassing realm of the "fictive We."

Finally, although every reading act is "public" in the sense that it occurs within a definite social and cultural context, it is easy to overlook this fact when discussing the literary reading act. Reading a novel seems such a private and individual activity—even if, as Stanley Fish contends, that individual reader merely applies the strategies of a larger interpretive community. The public or social dimensions of television "reading" are undeniable. The simultaneity of television broadcasts, with millions of sets receiving the same images at the same time, makes watching a television program a social phenomenon even if we are "alone" while we watch. The oceanic nature of television programming, its constant references to other texts, the close connections between television and other forms of textual production, all combine to plug any individual act of television viewing into a network of other viewings and other discourses, and to link us as viewers into the larger culture.

NOTES

1. Wolfgang Iser, *The Act of Reading: A Theory of Aesthetic Response* (Baltimore, Md.: Johns Hopkins University Press, 1978), pp. 3–5.

2. For those interested in the finer points of these debates and in the relationship of reader-oriented criticism to other types of analysis, a good introduction is provided by Robert C. Holub, *Reception Theory: A Critical Introduction* (London: Methuen, 1984). Holub concentrates on the strand of reader-oriented criticism developed in Germany, especially that associated with Wolfgang Iser, Hans Robert Jauss, and their colleagues at the University of Constance. William Ray considers Iser, Ingarden, Holland, Bleich, Culler, and Fish as a part of his more general survey of literary interpretations since phenomenology; see *Literary Meaning: From Phenomenology to Deconstruction* (London: Basil Blackwell, 1984). Terry Eagleton discusses reception theory from a Marxist perspective in *Literary Theory: An Introduction* (Minneapolis: University of Minnesota Press, 1983). Jonathan Culler, more an actor than an observer of disputes in reader-oriented criticism, relates the project of reader-oriented criticism to semiotics in *The Pursuit of Signs: Semiotics, Literature, and Deconstruction* (London: Routledge and Kegan Paul, 1981). Two collections of reader-oriented criticism have excellent introductory essays: Susan Suleiman and Inge Crossman, eds., *The Reader in the Text: Essays on Audience and Interpretation* (Princeton, N.J.: Princeton University Press, 1980); and Jane P. Tompkins, ed., *Reader-Response Criticism: From Formalism to Post-Structuralism* (Baltimore, Md.: Johns Hopkins University Press, 1980).

3. Georges Poulet, "Phenomenology of Reading," *New Literary History* 1 (October 1969): 56, quoted in Ray, *Literary Meaning*, p. 101.

4. See Roman Ingarden, *The Literary Work of Art*, trans. George C. Grabowicz (Evanston, Ill.: Northwestern University Press, 1973).

5. Iser, *The Act of Reading* (originally published in German as *Der Act Des Lesens: Theorie asthetischer Wirkung* [Munich: Wilhelm Fink, 1976]).

6. Consider, for example, the opening sentences of Fay Weldon's novel, *Female Friends*: "Understand, and forgive. It is what my mother taught me to

do, poor patient gentle Christian soul, and the discipline she herself practised, and the reason she died in poverty, alone and neglected" (*Female Friends* [London: Picador Books, 1977], p. 5). The first sentence would appear to be an injunction to "understand and forgive." But who should understand and forgive? Who is speaking these words? In what context are these admonitions meant? The following sentence answers these questions to some degree: the first-person narrator is the "speaker" of these words; they are precepts taught her by her mother. But it also asks another question: why are they the reason she "died in poverty, alone and neglected"? Obviously, we'll have to read on to find out. Perhaps the next sentence will answer this question, perhaps it will defer the answer until later, perhaps by saying nothing more about it in the following sentences the text will leave it to us to infer some causal relationship between her mother's philosophy of understanding and forgiving and her death. The last sentence of this very short chapter (only a few paragraphs) reads: "Such were Chloe's thoughts before she slept." Aha! Someone named Chloe "thought" what we have just read and she did so before going to sleep. Our understanding of all that we have read to this point is retrospectively altered by the information contained in this sentence. The blank paper at the bottom of that first page signals us to turn the page in the hope of discovering more about Chloe: Why is she thinking about her mother's precepts and death? And so forth. This question/answer/question chain continues throughout the novel.

7. The above discussion of Iser's theory of the reading act is largely based on chapter 5 of *The Act of Reading*, "Grasping a Text."

8. Iser, *The Act of Reading*, pp. 191–92.

9. Ibid., p. 192.

10. W. Daniel Wilson, "Readers in Texts," *PMLA* 96, no. 5 (October 1981): 848–63.

11. Umberto Eco, *The Role of the Reader* (Bloomington: Indiana University Press, 1977), p. 7.

12. Wayne C. Booth, *The Rhetoric of Fiction* (Chicago: University of Chicago Press, 1961), p. 138.

13. At the beginning of chapter 4, the narrator addresses the reader: "I know there are readers in the world, as well as many other good people in it, who are no readers at all—who find themselves ill at ease, unless they are let into the whole secret from first to last, of everything which concerns you." He then goes on for a few paragraphs as to the kind of background information the reader might desire to have at the beginning of a personal history. The narrator then says:

> To such, however, as do not choose to go so far back into these things, I can give no better advice than that they skip over the remaining part of this Chapter; for I declare beforehand, 'tis wrote only for the curious and inquisitive.
>
> —————————————Shut the door.—————————————
>
> I was begot in the night, betwixt the first Sunday and the first Monday in

the month of March . . . (Lawrence Sterne, *Tristram Shandy* [New York: New American Library Edition, 1962], p. 12).

Note that the relationship between this characterized reader and what we might presume to be the implied reader is left unclear and that as "real" readers we're not sure whether or not we want to be identified as the "reader" the narrator addresses. Are we among those readers "who find themselves ill at ease, unless they are let into the whole secret from first to last"? We are then forced to choose what kind of reader we are; we can take the narrator's advice and "shut the door" on the rest of the chapter, or we can align ourselves with the "curious and inquisitive" and read on.

14. Michele Hilmes discusses the "direct address" quality of television in "The Television Apparatus: Direct Address," *Journal of Film and Video* 37, no. 4 (Fall 1985): 27–36.

15. Robert Stam, "Television News and Its Spectator," in *Regarding Television—Critical Approaches: An Anthology*, ed. E. Ann Kaplan, American Film Institute Monograph Series, vol. 2 (Frederick, Md.: University Publications of America, 1983), p. 39.

16. Culler, *The Pursuit of Signs*, p. 50.

17. Holub, *Reception Theory*, pp. 100–101; Eagleton, *Literary Theory*, pp. 78–85.

18. Eagleton, *Literary Theory*, pp. 84–85.

19. Norman Holland, *5 Readers Reading* (New Haven, Conn.: Yale University Press, 1975), pp. 28–29.

20. Ray, *Literary Meaning*, pp. 67–68.

21. Stanley Fish, *Is There a Text in This Class?* (Cambridge, Mass.: Harvard University Press, 1980), p. 335.

22. Ibid., chapter 14.

23. Ibid., p. 370.

24. Culler, *The Pursuit of Signs*, p. 53.

25. Robert C. Allen, *Speaking of Soap Operas* (Chapel Hill: University of North Carolina Press, 1985), pp. 11–18.

26. Fish, *Is There a Text in This Class?* pp. 109–10.

27. Tony Bennett, "Text and Social Process: The Case of James Bond," *Screen Education* 41 (Winter/Spring 1982): 9.

28. Ibid., pp. 9–14.

29. The following discussion of the made-for-TV movie is based largely on Laurie Jane Schulze, "Text/Context/Cultural Activation: The Case of the Made-for-Television Movie," M.A. thesis, University of North Carolina at Chapel Hill, 1985.

30. Ibid., pp. 95–103.

31. Ibid., p. 171.

32. Bennett, "Text and Social Process," p. 14.

33. Raymond Williams, *Television: Technology and Cultural Form* (New York: Schocken Books, 1975), p. 90.

FOR FURTHER READING

As I suggested in the notes for this chapter, there are several good introductions to reader-response or reception theory. Robert C. Holub's *Reception Theory: A Critical Introduction* (London: Methuen, 1984) provides a critical overview of the work of the German reception theorists, particularly Wolfgang Iser and Hans Robert Jauss. Two excellent anthologies of reader-oriented literary criticism are Susan Suleiman and Inge Crossman, eds., *The Reader in the Text: Essays on Audience and Interpretation* (Princeton, N.J.: Princeton University Press, 1980); and Jane P. Tompkins, ed., *Reader-Response Criticism: From Formalism to Post-Structuralism* (Baltimore, Md.: Johns Hopkins University Press, 1980). Both have good introductory essays and bibliographies.

Several other works position reader-response criticism within a more general context of literary theory, among them: William Ray, *Literary Meaning: From Phenomenology to Deconstruction* (London: Basil Blackwell, 1984); Terry Eagleton, *Literary Theory: An Introduction* (Minneapolis: University of Minnesota Press, 1983); and Jonathan Culler, *The Pursuit of Signs: Semiotics, Literature, and Deconstruction* (London: Routledge and Kegan Paul, 1981).

Much of the analysis of soap opera structure in this chapter is based upon work by Iser and Jauss. Iser's approach is best laid out in *The Act of Reading: A Theory of Aesthetic Response* (Baltimore, Md.: Johns Hopkins University Press, 1978); see also *The Implied Reader: Patterns of Communication in Prose Fiction from Bunyan to Beckett* (Baltimore, Md.: Johns Hopkins University Press, 1974)—a collection of essays on various types of narrative prose. Jauss's more historical theory of reception is proposed in *Aesthetic Experience and Literary Hermeneutics* (Minneapolis: University of Minnesota Press, 1982), and *Toward an Aesthetic of Reception* (Minneapolis: University of Minnesota Press, 1982).

For examples of reader-oriented criticism written from perspectives other than that of Iser and Jauss, see David Bleich, *Subjective Criticism* (Baltimore, Md.: Johns Hopkins University Press, 1978); Harold Bloom, *The Anxiety of Influence: A Theory of Poetry* (Oxford: Oxford University Press, 1973); Umberto Eco, *The Role of the Reader* (Bloomington: Indiana University Press, 1977); Stanley Fish, *Is There a Text in This Class?* (Cambridge, Mass.: Harvard University Press, 1980); Norman Holland, *5 Readers Reading* (New Haven, Conn.: Yale University Press, 1975); Steven Mailloux, *Interpretive Conventions: The Reader in the Study of American Fiction* (Ithaca, N.Y.: Cornell University Press, 1982).

The relationship between viewer and television is examined in several of the essays included in E. Ann Kaplan, ed., *Regarding Television—Critical Approaches: An Anthology*, American Film Institute Monograph Series, vol. 2 (Frederick, Md.: University Publications of America, 1983), among them: Jane Feuer, "The Concept of Live Television: Ontology as Ideology," pp. 12–22; Robert Stam, "Television News and Its Spectator," pp. 23–43; and Sandy Flitterman, "The *Real* Soap Operas: TV Commercials," pp. 84–96. I attempt to develop what I call a "reader-oriented poetics" of the soap opera in *Speaking of Soap Operas* (Chapel Hill: University of North Carolina Press, 1985). Michele Hilmes discusses direct address in television in "The Television Apparatus: Direct Address," *Journal of Film and Video* 37, no. 4 (Fall 1985): 27–36.

Accounting for the reader or viewer's response has emerged as a major concern within British cultural studies. See David Morley and Charlotte Brunsdon's work on the news program *Nationwide* in Brunsdon and Morley, *Everyday Television: "Nationwide"* (London: British Film Institute, 1978); and Morley, *The "Nationwide" Audience: Structure and Decoding* (London: British Film Institute, 1980). Influenced by the British cultural studies approach to viewer studies is Ien Ang, *Watching "Dallas": Soap Opera and the Melodramatic Imagination*, trans. Della Couling (London: Methuen, 1985). John Fiske provides a more complete reading list at the end of his chapter on British cultural studies.

GENRE STUDY
AND TELEVISION
JANE FEUER

The term "genre" is simply the French word for type or kind. When it is used in literary, film, or television studies, however, it takes on a broader set of implications. The very use of the term implies that works of literature, films, and television programs can be categorized; they are not unique. Thus genre theory deals with the ways in which a work may be considered to belong to a class of related works. In many respects the closest analogy to this process would be taxonomy in the biological sciences. Taxonomy dissects the general category of "animal" into a system based on perceived similarity and difference according to certain distinctive features of the various phyla and species. As one literary critic has remarked, "biological classification is itself an explanatory system, which has been devised primarily to make sense of an otherwise disparate group of individuals and which is changed primarily in order to improve that sense. While robins and poems are obviously different, the attempt to make a reasoned sense similarly dominates their study."[1] In a similar way, literature may be divided into comedy, tragedy, and melodrama; Hollywood films into Westerns, musicals, and horror films; television programs into sitcoms, crime shows, and soap operas. Genre theory has the task both of making these divisions and of justifying the classifications once they have been made. Taxonomy has a similar task. However, the two part company when it comes to the question of aesthetic and cultural value. The purpose of taxonomy is not to determine which species are the most excellent examples of their type, or to illustrate the ways in which a species expresses cultural values, or to show how that species manipulates an audience, to mention varying goals of genre classification. But rather than discussing genre analysis as a whole, we should distinguish among the uses of the term for literature, film, and television.

Traditionally, the literary concept of genre has referred to broad categories of literature (such as comedy and tragedy) that tend not to be treated as historically or culturally specific manifestations. For example, Aristotle defined tragedy as an ideal type according to which any particular tragedy must be measured. Even though he drew upon the theater of his own society (classical Greece) for his models, Aristotle spoke of "tragedy" as a kind of overarching structure that informs individual works. Once the ideal structure was achieved, Aristotle implied, tragedy could then have its ideal impact on an audience. (In a similar way, although Hollywood film genres are constructed from actual films, the genre itself is frequently spoken of as an ideal set of traits that inform individual films. Thus, although many individual Westerns do not feature Indians, Indians remain a crucial generic element.) Drawing on Aristotle, the literary critic Northrop Frye attempted in the 1950s to further develop the idea of classifying literature into types and categories that he called "genres" and "modes." Frye commented that "the critical theory of genres is stuck precisely where Aristotle left it."² Frye attempted to further differentiate among types of literature. He classified fiction into *modes* according to the hero's power of action—either greater than ours, less than ours, or the same as ours—arriving at such categories as myth, romance, epic and tragedy, comedy, and realistic fiction according to the hero's relationship to the reader. Frye points out that over the last fifteen centuries these modes have shifted, so that, for example, the rise of the middle class introduces the "low mimetic" mode in which the hero is one of us (*AC*, pp. 33–35). As for *genres*, Frye distinguishes among drama, epic, and lyric on the basis of their "radical of presentation" (e.g., acted out, sung, read), viewing the distinction as a rhetorical one, with the genre being determined by the relationship between the poet and his public (*AC*, pp. 246–47).

We can see that the traditional literary view of genre would have only a limited application to film and television. The literary categories are very broad ones. Such literary types as drama and lyric, tragedy and comedy span numerous diverse works and numerous cultures and centuries. Film and television, however, are culturally specific and temporally limited. Instead of employing a broad category such as "comedy," we need to activate specific genres such as the "screwball comedy" (film) or the "situation comedy" (television), categories that may not correspond to or necessarily be subspecies of the literary genre of comedy. As we will see, attempts to measure the comic forms of mass media against the norms of drama are doomed to failure. At this point in the development of film genre theory, the concept has been applied most usefully to American film and television. Moreover, literary genres tend to be—to employ a distinction from Todorov—*theoretical* to a greater extent than do film and television genres, which tend to be *historical.*[3] The former are "deduced from a preexisting theory of literature," whereas the latter are "derived from observation of preexisting literary facts."[4] That is to say, some genres are accepted by the culture, whereas others are defined by critics.

Literary criticism, which has been around much longer than either film or television criticism, has described more genres from the theoretical or deductive perspective. Film and television criticism still tend to take their category names from current historical usage. For example, although Homer did not refer to his own work as an "epic" poem, both industry and critics employ the categories of "Western" and "sitcom." One of the goals of film and television genre criticism is to develop more theoretical models for these historical genres, not necessarily remaining satisfied with industrial or common-sense usage. Thus, in film genre study, the theoretical genre called *film noir* was constructed out of films formerly grouped under the historical labels "detective films," "gangster films," and "thrillers." Indeed, even melodramas such as *Mildred Pierce* were discovered to possess the stylistic traits of this newly created theoretical genre. TV studies is too new to have greatly differentiated between historical and theoretical genres; however, we are now attempting to redefine, if not reclassify, some of the received categories such as soap opera. Originally a derisive term used to condemn other forms of drama as being hopelessly "melodramatic," the term "soap opera" has been refined in a confrontation between such historical examples as the afternoon serial drama, the prime-time serials, and British soap operas. British "soaps," for example, cause us to question the equation of the term soap opera with the mode of melodrama, because their own mode might better be described as "social realism," possessing none of the exaggeration and heightened emotion and gestures of their American cousins. And the middle-class, plodding, woman-centered world of afternoon soaps bears little resemblance to the plutocratic worlds of *Dallas* and *Dynasty.*

Out of this confrontation emerges a new conceptualization of the genre, in which the continuing serial format is not necessarily equated with the descriptive term soap opera. Thus we can retain the *method* of the literary definition of genres without necessarily retaining their *content*. The literary concept of genre is based upon the idea, also common to biology, that by classifying literature according to some principle of coherence, we can arrive at a greater understanding of the structure and purpose of our object of study. Thus the taxonomist begins with already existing examples of the type. From these, he/she builds a conceptual model of the genre, then goes on to apply the model to other examples, constantly moving back and forth between theory and practice until the conceptual model appears to account for the phenomena under consideration. (Of course, this is a lot easier when the genre is already complete—not, as with television, when the genre is in a constant state of flux and redefinition.)

As Rick Altman points out, every corpus thus conceived reflects a particular methodology. The constitution of a generic corpus is not independent of or logically prior to the development of a methodology.[5] According to another literary critic, "What makes a genre 'good,' in other words, is its power to make the literary text 'good'—however that 'good' be presently defined by our audience."[6] Thus, what makes the popular artifacts of movies and television "good" may not correspond to the generic "good" of literary works.

It is due to their nature as artifacts of popular culture that films and television programs have been treated in a specific way in genre studies. Genre study in film has had a historically and culturally specific meaning. It has come to refer to the study of a particular kind of film— the mass-produced "formulas" of the Hollywood studio system. This concept of formula has been defined by John Cawelti:

> A formula is a conventional system for structuring cultural products. It can be distinguished from invented structures which are new ways of organizing works of art. Like the distinction between convention and invention, the distinction between formula and structure can be envisaged as a continuum between the two poles; one pole is that of a completely conventional structure of conventions—an episode of the Lone Ranger or one of the Tarzan books comes close to this pole; the other end of the continuum is a completely original structure which orders inventions—*Finnegan's Wake* is perhaps the ultimate example.[7]

In this way, the concept of genre stems from a conception of film as an industrial product. That is, the particular economic organization of the film industry led to a kind of product standardization antithetical to the literary concept of an authored work. Thus, film genre study has always referred back to the capitalist mode of production; it is potentially more materialist than other ways of categorizing the products of mass culture.

Within the institution of film criticism, however, the concept of genre was initially employed to condemn mass-produced narratives such as Hollywood studio films for their lack of originality. It was assumed that genre films could not have any artistic merit, because they were not original works and because they were not authored works. These standards of evaluation are based upon a Romantic theory of art, one that places the highest value on the concepts of originality, personal creativity, and the idea of the individual artist as genius. Ironically, it was through an attempt to establish a Romantic, author-centered model for film that the concept of genre began to take on a more positive meaning in film criticism. The *auteur* policy attempted to reconceptualize the anonymous products of the Hollywood assembly line as the creations of individual artists, assumed to be the directors of the films. The author was constructed by attributing unity—whether stylistic or thematic or both—to those films possessing the signature of certain directors. One would think that the *auteur* approach would have further invalidated genre criticism. However, it was discovered that certain authors expressed themselves most fully within a particular genre—John Ford in the Western or Vincente Minnelli in the musical. In some sense, then, the genre provided a field in which the force of individual creativity could play itself out. Some viewed the genre as a constraint on complete originality and self-expression, but others, following a more classical or mimetic theory of art, felt that these constraints were in fact productive to the creative expression of the author. Thus genre study evolved within film studies as a reaction against the Romantic bias of *auteur* criticism.

When film studies turned toward semiotics and ideological criticism, the idea of the genre as a threshold or horizon for individual expression gave way to an interest in the genres themselves as system and structure. Thomas Schatz has referred to the semiotic interest in genre as "the language analogy." He says that genre can be studied as a formalized sign system whose rules have been assimilated (often unconsciously) through cultural consensus. Following Levi-Strauss, Schatz views genres as cultural problem-solving operations. He distinguishes between a deep structure that he calls "film genre" and a surface structure that he calls the "genre film." The genre film is the individual instance, the individual utterance or speech act (*parole*). The film genre is more like a grammar (*langue*), that is, a system for conventional usage. According to Schatz, the film genre represents a tacit contract between the motion-picture industry and the audience, whereas the genre film represents an event that honors that contract. According to this linguistic view, a film genre is both a static and a dynamic system. However, unlike language, individual utterances do have the capacity to change the rules.[8] To take an example from television, by introducing overt political content, *All in the Family* altered the grammar of the television sitcom.

The language analogy sees an active but indirect participatory role for the audience in this process of genre construction. For the industrial arts, the concept of genre can bring into play (1) the system of production, (2) structural analysis of the text, and (3) the reception process, with the audience conceived as an interpretive community. Rick Altman relates the concept of genre to that of the interpretive community. For him, the genre serves to limit the free play of signification and to restrict semiosis. The genre, that is to say, usurps the function of an interpretive community by providing a context for interpreting the films and by naming a specific set of intertexts according to which a new film must be read. The genre limits the field of play of the interpretive community. Altman sees this as an ideological project because it is an attempt to control the audience's reaction by providing an interpretive context. Genres are thus not neutral categories, but, rather, they are ideological constructs that provide and enforce a pre-reading.[9] In a similar way, Steve Neale sees genres as part of the dominant cinema's "mental machinery," not just as properties possessed by texts. Neale defines genres as "systems of orientations, expectations, and conventions that circulate between industry, text and subject." Any one genre, then, is both a "coherent and systematic body of film texts" and a coherent and systematic set of expectations. Neale agrees with Altman that genres limit the possibilities of meaning, both exploiting and containing the diversity of mainstream cinema.[10] Drawing upon Altman and Neale, we can conclude that each theoretical genre is a construct of an analyst. The methodology that the analyst brings to bear upon the texts determines the way in which that analyst will construct the genre. Thus genres are made, not born. The coherence is provided in the process of construction, and a genre is ultimately an abstract conception rather than something that exists empirically in the world.

Thus we can distinguish a number of different reasons why the concept of genre has figured in both popular and critical discourses as an "instrument for the regulation of difference."[11] From the television industry's point of view, unlimited originality of programming would be a disaster because it could not assure the delivery of the weekly audience, as do the episodic series and continuing serial. In this sense, television takes to an extreme the film industry's reliance upon formulas in order to predict audience popularity. For the audience—as members of various interpretive communities for American mass culture—genre assures the interpretability of the text. Through repetition, the cultural "deep structure" of a film genre "seeps to the surface." The audience—without conscious awareness—thus continually rehearses basic cultural contradictions that cannot be resolved within the existing socioeconomic system outside of the text: law and order versus the idea of individual success (the gangster genre); nature versus culture (the Western); the work ethic versus the pleasure principle (the musical).

The approaches to genre we have discussed might be summarized under three labels—the aesthetic, the ritual, and the ideological approaches. Although in practice these are not absolutely distinct, in general we can use them to distinguish among different approaches that have been taken toward film and television genres. The *aesthetic approach* includes all attempts to define genre in terms of a system of conventions that permits artistic expression, especially involving individual authorship. The aesthetic approach also includes attempts to assess whether an individual work fulfills or transcends its genre. The *ritual approach* sees genre as an exchange between industry and audience, an exchange through which a culture speaks to itself. Horace Newcomb and Paul Hirsch refer to television as a "cultural forum" involving the negotiation of shared beliefs and values, and helping to maintain and rejuvenate the social order as well as assisting it in adapting to change.[12] Most approaches based on the language analogy take the ritual view. The *ideological approach* views genre as an instrument of control. At the industrial level, genres assure the advertisers of an audience for their messages. At the textual level, genres are ideological insofar as they serve to reproduce the dominant ideology of the capitalist system. The genre positions the interpretive community in such a way as to naturalize the dominant ideologies expressed in the text. However, some ideological critics allow for constant conflict and contradiction in the reproduction of ideology, as the ruling ideas attempt to secure hegemony. A more reader-oriented ideological model would allow for the production of meanings by the viewer as well. Thus recent approaches to genre have attempted to combine the insights of the ritual approach with those of the ideological approach. According to Rick Altman, "because the public doesn't want to know that it's being manipulated, the successful ritual/ideological 'fit,' is almost always one that disguises Hollywood's potential for manipulation while playing up its capacity for entertainment. . . . The successful genre owes its success not alone to its reflection of an audience ideal, nor solely to its status as apology for the Hollywood enterprise, but to its ability to carry out both functions simultaneously." [13]

THE SITUATION COMEDY

As an example of the generic approach to television analysis, I have chosen to discuss the most basic program format known to the medium—the situation comedy. In general, television taxonomy has not yet advanced to the point where a clear distinction between historical and theoretical genres has emerged. Thus all TV genres in some sense remain historical genres, those defined by a consensus between the industry, *TV Guide,* and the viewing audience. The sitcom is no exception. We are all capable of identifying its salient features: the half-hour format, the basis in humor, the "problem of the week" that causes the hilarious situation and that will be resolved so that a new episode may come on next week.

Nevertheless, different methodologies for defining the genre have produced different notions of the sitcom as genre. I will discuss the ways in which three critics have approached the genre in order to demonstrate that each has constructed a *different* genre called the sitcom. David Grote takes a literary approach to the genre and finds that it lacks development of any kind, serving merely to reassert the status quo. Horace Newcomb also finds the genre limited in its capacity for ambiguity, development, and the ability to challenge our values; however, because he takes a ritual view, he does see the genre as basic to an understanding of the reassurance the television medium provides for its audience. David Marc appears to believe that certain authors can make the sitcom form into social satire; his would represent an aesthetic approach. Finally, my own approach will be a synthetic one, viewing the sitcom as a genre that did develop, for historical reasons, in the direction of the continuing serial.

The most literary—and consequently the most negative—view of the television sitcom is taken by David Grote. According to Grote, television has completely rejected the type of comic plot that has dominated the comedic tradition from Greek and Roman times—a type that, following Northrop Frye, he calls "new comedy." In the tradition of new comedy, a very basic arrangement of plot and character has predominated. In it, a young man's desire for a young woman meets with resistance, usually by her father, but before the end of the play, a plot reversal enables the boy to get the girl. This is the plot of Greek new comedy, which can be dated back to 317 B.C., but it is also the plot of Shakespeare's comedies, of Hollywood romantic comedies, and of many musical comedies. While few would dispute the longevity of this plot paradigm, many might question Grote's next step, which is to make a sweeping historical generalization about the social meaning of new comedy and then to use that generalization to disparage the sitcom as a new form of comedy that rejects that social meaning. According to Grote, the comic plot is social in nature because the forces that keep the lovers apart always represent social authority. The resistance of the young lovers to the parental figure thus represents a threat to power, authority, and stability, because, according to Grote, in this type of comedy father *never* knows best.

At the end of the traditional ("new") comedy, there is a celebration—usually the wedding of the young people—at which the father is invited back in. The authority figure actually admits that he was wrong and the rebellious children right. The basic comic plot uses the young couple's union to symbolize the promise of the future, guaranteeing the possibility of personal change and, with it, social change. In this way Grote assumes that the basic comedy plot has held the same meaning in different cultures and throughout history, thus conceptualizing the genre as an ideal type with a single, ahistorical, acultural meaning. His next step is even more universalizing: he claims that the TV sitcom completely rejects both the form and the meaning of this traditional comic plot, thus symbolizing the "end of comedy" as a progressive social force.

Grote bases his static conception of the sitcom form on its nature as an episodic series, that is, a program with continuing characters but with a new plot (situation) each week. Thus, no matter what happens, the basic situation can't change. From this, Grote generalizes that the sitcom resists the change of the traditional comic plot and indeed resists change of any kind:

> The situation comedy as it has evolved on American television has rejected more than the traditional comedy plot. Not only does boy not pursue and capture girl, he does not pursue *anything*. The principal fundamental situation of the situation comedy is that things do not change. No new society occurs at the end. The only end is death, for characters as well as for the situation itself, the precise opposite of the rebirth and new life promised in the celebrations of the traditional comedy. The series may come on every week for no more reason than that it is convenient for the network and the sponsors, but the messages that accompany those weekly appearances are the messages of defense, of protection, of the impossibility of progress or any other positive change. . . . That such a change occurred is curious, but that such a change occurred in the largest mass medium known to man, in the most progressive and changeable society in Western history, and was immensely popular, is almost incredible. Everything the traditional comedy stood for, at every level of art, psychology, philosophy, and myth, has been overthrown in this New Comedy of American television.[14]

I have chosen to discuss Grote's "construction" of the genre not because I think it is the construction that does the most "good" for the texts, but rather because I think it takes to an extreme a very common view that the TV sitcom is by nature a conservative and static form. The goal of the sitcom, according to Grote, is to reaffirm the stability of the family as an institution. Thus Grote moves, as would any genre analyst, from an identification of the formal features of the text (in this case the nature of the episodic series and the fact that each episode returns to the equilibrium with which it began), to a generalization about the meaning of these features (they represent a rejection of change of any kind) to a social, cultural, political, or aesthetic interpretation of the genre (the sitcom represents the end of the progressive potential of the traditional comic plot). If we accept Grote's premises, his conclusions are not illogical. However, his entire argument depends on an acceptance of his belief that after centuries of progressiveness, the meaning of comedy suddenly shifted to a regressive one for no other reason than that the television medium has transformed history. Many would find this difficult to accept as an historical explanation.

Yet even the more complex "ritual" view of the genre constructed by Horace Newcomb bases its model for the genre on the formal qualities of the episodic series. For Newcomb (writing in 1974, and thus without full knowledge of the developments in the genre during the 1970s), the sitcom formula provides a paradigm for what occurs in more complex program types and provides a model of a television formula in that "its rigid structure is so apparent." [15] The situation is "the funny thing that will happen this week." Next week there will be a new situation entirely independent of what happened this week. The situation develops through complication and confusion usually involving human error. There is no plot development and no exploration of ideas or of conflict: "The only movement is toward the alleviation of the complication and the reduction of confusion" (*TV*, p. 34). Thus Newcomb sees the sitcom as providing a simple and reassuring problem/solution formula. The audience is reassured, not challenged by choice or ambiguity, nor are we forced to reexamine our values. When the sitcom shifts its meaning away from situations and toward persons, we find ourselves in a slightly different formula, that of the "domestic comedy," says Newcomb. Newcomb defines the domestic comedy as one in which the problems are mental and emotional; there is a deep sense of personal love among members of the family and a belief in the family—however that may be defined—as a supportive group. Although, as with the sitcom, the outcome is never in doubt, for the domestic comedy "it is also true that there is more room for ambiguity and complexity, admittedly of a minimal sort. Characters do seem to change because of what happens to them in the problem-solving process. Usually they 'learn' something about human nature" (*TV*, p. 53). Newcomb goes on to point out that the form of the domestic comedy may expand when problems encountered by the family become socially or politically significant (as in *All in the Family* or *M*A*S*H*).

Newcomb thus constructs the sitcom as the most "basic" of the television genres in the sense that it is the furthest from "real world" problems such as are encountered in crime shows, and from real world forms and value conflicts such as are encountered in soap operas. It is, in a sense, formula for formula's sake; the very ritualistic simplicity of the problem/solution format gives us a comforting feeling of security as to the cultural status quo. Newcomb thus constructs a ritual view of the genre, but a ritual view based upon an essentially static conception of the episodic series such as had informed Grote's more universalized and literary account. Newcomb's major interpretation comes in the equation of the form with a cultural meaning of stability and reassurance. For it is equally possible to view the static nature of the sitcom form as having the potential to challenge our received norms and values.

This is the position that David Marc appears to take in his chapter on "The Situation Comedy of Paul Henning."[16] Marc attributes the subversive potential of sitcoms such as *The Beverly Hillbillies* to the presence of an author—in this case, the producer Paul Henning—thus making his an aesthetic conception of genre (i.e., an author can work in a banal genre such as the sitcom and transform it into an individual statement). Nevertheless, the argument for the subversive potential of the static sitcom form need not depend upon the aesthetic conception but may be seen to lie in the ideology of the genre itself, quite apart from what a particular author may choose to do with it.

For Grote and, to a lesser extent, Newcomb, *The Beverly Hillbillies* would qualify as a basic episodic sitcom that endlessly replays the theme of the virtue of plain values and the rejection of materialism. For Marc, the show is a brilliant caricature of cultural values and conflicts, in its way as much of a social critique as *All in the Family*. We can find the theme of the backwoodsman versus the city slicker in American folklore and in other television genres as well (the family dramas of *Little House on the Prairie* and *The Waltons* frequently feature this theme). In the sitcom, however, the theme is treated comically, giving it a satirical potential.

Marc would agree with Grote and Newcomb that on *The Beverly Hillbillies,* the plots never develop very far: the Clampetts never adjust to life in Beverly Hills; the family is never accepted by their neighbors; Elly May never marries; Granny never gives up her mountain ways. But Marc does not evaluate this lack of development in a negative light. Rather, he sees the Henning sitcoms as a departure from the formula of the 1950s sitcom. Unlike Newcomb's domestic comedy, in Henning sitcoms the individual crisis of a family member does not provide us with the weekly situation. We don't identify emotionally with the Clampett's problems, as we would in a program with more psychological development of characters, so that instead, *The Beverly Hill-billies* provides us with an almost pure cultural conflict. Marc says that we are invited to test our own cultural assumptions because "the antagonists are cultures" and the characters "charged cultural entities." He concludes that Paul Henning's *The Beverly Hillbillies*, while not satire per se, is nonetheless a "nihilistic caricature of modern life."

Thus Marc differs from Grote and Newcomb not over their description of the sitcom's lack of plot and character development, but rather over their interpretation of what this essentially static genre means. For Grote, it means that the sitcom is inferior to the dominant literary form of comedy; for Newcomb, that it aids in the restoration and maintenance of society. For Marc, it would seem to mean something entirely different: he implies that Henning's comic treatment may be more socially satirical than the expansive form of domestic comedy that accommodates social and political issues (the Norman Lear sitcoms of the early- to mid-1970s being the epitome of this type). In this way the static sitcom structure *can* explore ideas and challenge dominant cultural values, and it is able to do so precisely because it does not allow our individualistic identification with well-developed characters to get in the way. If we follow out the logic of this point of view, it could lead to the conclusion that *The Beverly Hillbillies* was more of a social satire than *All in the Family*, in which our identification with the more well-rounded Archie Bunker was likely to outweigh the positive liberal benefits of the show's intended satire of his racist beliefs.

Although all three models represent useful individual constructions of a television genre, none seems to me to account for the role of the interpretive community in the construction of a genre or the role of history in generic "evolution." In fact, one of the dangers of a generic approach is a built-in tendency to structuralize the model in such a way that it is impossible to explain changes or to see a genre as a dynamic model. The basis of much genre theory in the language analogy tends to remove it from history as well and to emphasize structure over development. When applied to the television medium, this danger is even greater, for we already have cultural preconceptions as to the "sameness" of television programming, that is, "if you've seen one sitcom, you've seen them all." The impression of continuity over difference intensifies when television is evaluated according to literary conceptions of genre, with their centuries of evolution, or even according to the half-century span of film genres such as the Western. I would argue, however, that the sitcom has "evolved" in its brief lifetime, in the sense that it has gone through some structural shifts and has modulated the episodic series in the direction of the continuing serial. This is not to say that the genre has "progressed" or become "better," but rather that it has become different. Unlike Grote, I think the changes need to be explained, but I also think that explaining such changes must be part of a complex construction of the genre.

As an example of how I would construct the genre, let's trace the development of the situation comedy from the late 1960s to the present. In order to do this, we have to take into account developments in the industry, social and cultural history, and developments more or less internal to the genre.[17] A common explanation for the move away from the "rural" sitcoms of the late 1960s and toward the social and political domestic comedy of the early- to mid-seventies is that the audience "felt a need" for a more sophisticated conception of the genre. Then, in the mid-seventies, they wanted the "mindless" teen-oriented sitcoms. In the 1980s, they desire family warmth, which signifies a return to the wholesome domestic comedies of the 1950s. The explanation of generic evolution/programming trends according to an assumed "need" on the part of the interpretive community is the most common way in which industry observers and participants construct TV genres. As an historical construct, it is worthy of analysis in itself (why this construction and not another?); as a theoretical construct, however, it begs the question. The concept of audience "need" is a substitute for an explanation of shifts in a culture, in an industry, and in a narrative form; in itself it does not explain anything. In at least one instance—the emergence of the MTM and Lear sitcoms in 1971—it can be demonstrated that what changed was *not* the demands of any empirical audience, but rather the industry's own construction of network television's interpretive community. Whereas in the era of the Paul Henning "hayseed" sitcoms the industry had conceptualized the audience as an aggregate or mass, it was now reconceptualizing the audience as a differentiated mass possessing identifiable demographic characteristics. This also caused the industry to redefine the measure of the popularity of a particular genre or program. Now "popularity" came to mean high ratings with the eighteen- to forty-nine-year-old urban dweller, rather than popularity with the older, rural audience that had kept the Paul Henning sitcoms on the air throughout the 1960s. Later, the industry refined its model audience once again. During the "Silverman years" of the mid- to late-1970s, the audience for sitcoms was defined as mindless teenagers; hence *Three's Company, Happy Days,* and *Laverne and Shirley.* In the 1980s, the desirable audience—at least for the NBC network—became the high-consuming 'yuppie' audience, thus defining the popularity of shows such as *Cheers* and *Family Ties.*

Of course, the audience itself no doubt changed from the late 1960s to the mid-1980s—specifically, the baby boomers matured during this period. And, of course, cultural changes no doubt influenced the generic shifts in the sitcom. But they did not directly cause the genre to change. It seems clear that the industry acted as an intermediary factor, in that it was continually redefining the audience for its own ends. An interesting question to pose would be: what caused the industry to redefine the audience at certain points, and to what extent did this really correspond to material changes in the culture? To further complicate the causality, the sitcom itself was responding to changes in other television genres—specifically, to what I would label the serialization of American television, throughout the 1970s.

Thus the sitcom, around 1970, shifted away from the "one dramatic conflict series" model of *The Beverly Hillbillies* and toward an expanded conception of the domestic comedy.[18] This was not necessarily as abrupt a shift as it now seems; earlier programs such as *The Dick van Dyke Show* (1961–66) had prepared the way for a reconceptualization of the domestic comedy in the direction of the home/office blend that would characterize the MTM sitcoms of the 1970s. Specifically, in the early seventies the sitcom was developed by two independent production companies (themselves responses to industrial changes): MTM Enterprises, which produced *The Mary Tyler Moore Show, The Bob Newhart Show, Rhoda,* and others; and Norman Lear's Tandem Productions, which produced *All in the Family, Maude, The Jeffersons,* and others. The aesthetic view comes into play here in the sense that the independent production companies encouraged the development of the writer/producer as a crucial creative component in the development of the new form of domestic comedy. (Of course, the emergence of the writer/producer was itself dependent upon cultural and industrial factors.)

We might say that the MTM and Lear sitcoms transformed the situation/domestic comedy by adapting the problems encountered by family members either in the direction of social and political issues (Lear) or in the direction of "lifestyle" issues (MTM). Thus the Bunker family had to deal with problems caused by blacks moving into the neighborhood, whereas Mary and Rhoda had to deal with problems caused by their being representatives of a new type of woman—working, single, independent, and confused. The basic problem/solution format of the sitcom did not change. Rather, the nature of the problems shifted and the conception of character held by the sitcom genre altered.

The Lear sitcoms were more influential in shifting the terrain of the characters' problems, whereas the MTM sitcoms were more influential in altering the conception of character. We have already seen that the assumed apolitical nature of the pre-1970s sitcom is called into question by new constructions of the genre through readings of programs such as *The Beverly Hillbillies*. Such readings assume that over the years the cultural conflict endlessly repeated in that show must have had some impact on the audience, however unconsciously that impact was assimilated. Nevertheless, the Lear sitcoms introduced an overtly political agenda into the genre. But it was in their conception of character that the "new wave" sitcoms of the 1970s most markedly altered the "grammar" of the formula.

The new domestic comedies introduced a limited but significant concept of character development into the genre. Although all comic characters are of necessity stereotyped (i.e., they possess a limited number of traits compared to actual individuals), the new sitcom characters were less stereotyped than their predecessors, especially in the MTM "lifestyle" variety. If the hillbillies never adapted to modern life, the same could not be said for Mary, who began her show by moving from a small town to Minneapolis in order to start a career. If previous characters in domestic comedy learned a little from experience, Mary learned a lot. Over the seven years the program was on the air, she became more assertive, more her own person. Similarly, Rhoda went from single womanhood to marriage to divorcée status within the span of her own series, each experience registering on the character and deepening our sense of her life experience. As television characters, the MTM women appeared to possess a complexity previously unknown to the genre. As both the nation and the industry became more conservative in the mid-1970s, the grammatical innovations of the Lear programs appeared passé, as political relevance faded from the sitcom's repertoire. But MTM's "character comedy" survived the transition from the new wave sitcoms of the early 1970s to the Silverman programs of the mid- to late-1970s. Then, under the impetus of an overall serialization and "yuppification" of American television in the 1980s, the MTM sitcom emerged as the dominant form of the genre.

The idea of character development inevitably moves a genre based on the episodic series model toward the continuing serial form. This is what occurred, for example, when Rhoda's wedding and subsequent divorce gave the episodes of that sitcom a continuing plot line and character continuity. But character development is also a quality prized by the upscale audience, which tends to have a more literary standard of value. We have already seen that the idea of character depth and development does not necessarily make for "better" or even for more sophisticated programming. To value "character comedy" over other comic techniques is to take up an ideological position, to construct the genre in a particular way and to value it for a kind of depth that some would construe as ideologically conservative. According to certain Marxist analyses of art (in particular, Bertolt Brecht's concept of the epic theater), flat characters are more politically progressive because they take us away from our identification with the characters and force us to think about how the play is constructed. According to this view, character complexity and development is merely a representation of bourgeois values. We have already seen a version of the Brechtian position in the argument that the concept of character in *The Beverly Hillbillies* is more socially critical than the concept of character in *All in the Family* or in *Cheers*. And, finally, character growth and development over time, along with an awareness of its own past, has always characterized the continuing serial, which, with the growing popularity of daytime serials in the late 1970s and the emergence of the prime-time serial genre with *Dallas*, finally emerged as a new narrative paradigm for generic television. The evolving sitcom had helped to prepare the way for the growth of serial drama; reciprocally, serialization gave a new grammar to the upscale comedies of the eighties. *Cheers* is a good example of the eighties sitcom designed to capture the upscale demographic audience. Sam and Diane develop from season to season. After their torrid affair of the second season, and their breakup in the third season, an episode in the fourth season harks back to the past. Thinking they are about to perish in an airplane crash, Sam confesses that he should have married Diane. That same season, they almost rekindle their lust for one another. This gives their relationship a sense of development and the series a sense of history.

At the same time, another "lifestyle" sitcom, *The Cosby Show*, returns us to the father-knows-best world of the 1950s domestic comedy, a world from which class and racial conflict are once again absent. The element of struggle in the Lear sitcoms would seem to have been put aside. Yet this absence has a different ideological motivation in the 1980s. The implication is that racial and economic equality have already been achieved, whereas in the fifties they were not yet seen as problems worthy of incorporation into the ideology of the domestic comedy. In this manner the sitcom genre develops unevenly, with different sitcoms operating at different points in the genre's ideological transformation.

The argument just made might lead a genre analyst to conclude that the sitcom does not fit theories of generic evolution developed for Hollywood film genres. According to the most teleological version of the theory of generic evolution, a genre begins with a naive version of its particular cultural mythology, then develops toward an increasingly self-conscious awareness of its own myths and conventions. It is implied that the genre is also progressing toward a higher version of its type. Although it is possible to construct the TV sitcom according to this evolutionary model, one could equally argue that the sitcom has gone through repeated cycles of regression to earlier incarnations, as exemplified by the cycle of "mindless" teen comedies of the 1970s and by the return to the traditional domestic comedy in the mid-eighties. Another theory of film genre development argues that after a period of experimentation, a film genre settles upon a classical "syntax" that later dissolves back into a random collection of traits, now used to deconstruct the genre.[19] This theory does not attempt to judge the value of any stage of generic development, nor does it see a genre as necessarily progressing toward a more perfect form. Yet it is difficult to see how this theory would apply to the TV sitcom either. There have been sitcoms that reflect back upon earlier ones (elsewhere I have argued that *Buffalo Bill* represented an inversion of the idea of the family of coworkers epitomized by *The Mary Tyler Moore Show*).[20] Yet even when it is possible to identify a period during which a stable "syntax" prevailed in the genre—such as the MTM/Lear hegemony of the 1970s—it is not as easy to point to a movement toward ever-greater self-reflexivity in a genre such as the sitcom. Rather, it would seem that the genre has gone through a series of transformations, some of which returned it to earlier versions of its own paradigm. Indeed, when

U.S. network television took on a greater self-reflexivity in the late 1970s with programs such as *Saturday Night Live* and *SCTV Comedy Network,* self-consciousness tended to emerge across genres rather than within them.

The problems involved in applying the theory of film genre evolution to television should remind us that genre theory as a whole might work better for film than for TV. Film genres really were mechanisms for the regulation of difference. The genre organized large numbers of individual works into a coherent system that could be recognized by the interpretive community. Television has always employed standard program types, but arguably this has not been the main principle of coherence for the medium. Television programs do not operate as discrete texts to the same extent as did movies; the property of "flow" blends one program unit into another, and programs are regularly "interrupted" by ads and promos. Thus many critics have argued that perhaps the unit of coherence for television is at a level larger than the program and different from the genre—for example, an evening's viewing on a particular network or all the possible combinations of programs a viewer could sample on an evening.

Theories of the evolution of film genres have argued that genres such as the Western and the musical develop by recombining and commenting upon earlier instances of their *own* genre. Of course, it was not uncommon during the Hollywood studio era (and it is even more common in contemporary Hollywood films that no longer exhibit the distinct genre boundaries of yore) for new instances to develop out of the recombination of previous genres. After all, one of the best-known "musicals" ever—*Oklahoma!*—could be considered a musical Western. But it is arguable that Hollywood genres had a greater tendency to draw upon their own predecessors, thus keeping generic boundaries relatively distinct and enabling them to serve an ideological function for the interpretive community as they recombined in ever more complex ways. Television genres, on the other hand, appear to have a greater tendency to recombine *across* genre lines. For instance, *Hill Street Blues* might be described as a crime show soap opera documentary that resembles the medical show *St. Elsewhere* far more than other crime shows or soap operas. And there exists an entire TV "genre"—the late-night comedy show—whose raison d'être appears to be to comment upon the whole range of television genres. This greater horizontal recombination also points to the limitations of the typically vertical consideration of the development of film genres. The genre approach has its limits in the process of constructing an understanding of the medium. Yet, as this chapter has tried to demonstrate, it also has its virtues.

NOTES

1. Adena Rosmarin, *The Power of Genre* (Minneapolis: University of Minnesota Press, 1985), p. 167.
2. Northrop Frye, *Anatomy of Criticism* (Princeton, N.J.: Princeton University Press, 1957), p. 13. Subsequent references will be cited in the text.
3. Tzvetan Todorov, *The Fantastic: A Structural Approach to a Literary Genre* (Cleveland, Ohio: Case Western Reserve University Press, 1975), pp. 13–14.
4. Rosmarin, *Power of Genre*, p. 26.
5. Rick Altman, *The American Film Musical* (Bloomington: Indiana University Press, 1987).
6. Rosmarin, *Power of Genre*, p. 49.
7. John Cawelti, *The Six Gun Mystique* (Bowling Green, Ohio: Bowling Green University Popular Press, 1970), p. 29.
8. Thomas Schatz, *Hollywood Genres* (New York: Random House, 1981), pp. 15–20.
9. Altman, *American Film Musical*.
10. Steve Neale, *Genre* (London: British Film Institute, 1980), p. 20.
11. Ibid., p. 19.
12. Horace M. Newcomb and Paul M. Hirsch, "Television as a Cultural Forum: Implications for Research," *Quarterly Review of Film Studies* 8, no. 3 (1983): 45–55.
13. Rick Altman, "A Semantic/Syntactic Approach to Film Genres," *Cinema Journal* 23, no. 3 (Spring 1984): 14–15.
14. David Grote, *The End of Comedy: The Sit-Com and the Comedic Tradition* (Hamden, Conn.: Shoestring Press, 1983), p. 105.
15. Horace Newcomb, *TV: The Most Popular Art* (New York: Anchor, 1974), p. 28. Subsequent references will be cited in the text.
16. David Marc, *Demographic Vistas: Television in American Culture* (Philadelphia: University of Pennsylvania Press, 1984), pp. 39–63.
17. For a more extensive discussion of the sitcom, see Jane Feuer et al., *MTM: Quality Television* (London: British Film Institute, 1984); and Jane Feuer, "Narrative Form in Television," in *High Theory, Low Culture*, ed. Colin MacCabe (Manchester, Eng.: Manchester University Press, 1986), pp. 101–14.
18. The phrase "one dramatic conflict series" is from Marc, *Demographic Vistas*, p. 62.
19. Altman, "Semantic/Syntactic Approach."
20. See "The MTM Style," in Feuer et al., *MTM*, pp. 52–56.

FOR FURTHER READING

This chapter has emphasized traditional literary conceptualizations of genre. For a contemporary Marxist view of literary genres (at an advanced level), see Fredric Jameson, *The Political Unconsciousness: Narrative As a Socially Symbolic Act* (Ithaca, N.Y.: Cornell University Press, 1981), particularly ch. 2, "Magical Narratives: On the Dialectical Use of Genre Criticism."

A readable work on film genre theory that also contains detailed critical and historical analysis of particular genres is Thomas Schatz, *Hollywood Genres* (New York: Random House, 1981). The most complete treatise to date on film genre theory is Rick Altman, *The American Film Musical* (Bloomington: Indiana University Press, 1987), which also contains a complete examination of the musical genre. Less readable, but more in the tradition of continental theory, is Steve Neale, *Genre* (London: British Film Institute, 1980).

The major work to date on television genres, and also a highly accessible one, is still Horace Newcomb, *TV: The Most Popular Art* (New York: Anchor, 1974). A number of books and articles deal with television genres, although not necessarily from a "genre studies" perspective. An article that does deal specifically with the concept of genre, employing Neale's theory of genre, is Paul Attallah, "The Unworthy Discourse: Situation Comedy in Television," in *Interpreting Television: Current Research Perspectives*, ed. Willard D. Rowland, Jr., and Bruce Watkins (Beverly Hills, Calif.: Sage, 1984), pp. 222–49.

A theory of the evolution of film genres is offered in Jane Feuer, *The Hollywood Musical* (London: Macmillan, 1982), especially ch. 5. This type of theory is tested upon television genres in Mimi White, "Television Genres: Intertextuality," *Journal of Film and Video* 37, no. 3 (1985): 41–47.

IDEOLOGICAL ANALYSIS AND TELEVISION

MIMI WHITE

THE CONTEXT OF IDEOLOGICAL CRITICISM

A recent commercial campaign on television features a male actor addressing the camera. "I'm not a doctor," he says, "but I play one on TV." This introduction furnishes the ground for his pitch, promoting Vicks cough medicines as an "adult formula." Varying versions of the ad have appeared, each featuring a different spokesman for the product. When it is broadcast within the flow of television, the commercial activates a range of assumptions, at once obvious and unspoken, about the medium in general and the normative expectations that inform its functioning:

1) In the context of American television, advertising is normal. It is recognized by viewers as the source of station/network income, and expected within the course of programming. Although it is beyond the purview of this chapter to explore the historical and institutional bases of this practice, it is crucial to underscore that the integral presence of commercials is taken as a given, regulating the rhythm and patterning of programs and viewing. The viewer, familiar with the regularity of commercials as an integral feature of textual flow, is addressed as a potential consumer.

2) The commercial is for a particular brand of cough medicine, one among many different brands that are advertised on television. They all claim to offer the best remedy for a particular common ailment. They vary not in what they do—reduce cough and flu symptoms—but in how they structure their appeal to potential consumers. A careful balance must be maintained within the ad between similarity and difference: this product is one version of a range of similar products, but is differentiated by brand name and by the narrative rhetoric of a particular campaign.

3) The product is itself divided within the ad. There are different versions of it, each serving as a remedy for a specific combination of cough and flu symptoms. This internal division—they all share a brand name, but have a different combination of ingredients—provides an image of bountiful inclusivity. For coughs and flu one need not look beyond this particular brand name product. However, this division is not unique to this product. Other manufacturers of patent medicines offer a similar choice of three or four concoctions of medicine, each a different color, to alleviate a range of symptoms. The balance of similarity and difference between brands—an image of plenitude and free market choice—is thus duplicated with the ads for each brand name.

4) The persona of the spokesman is established as authoritative through a structure of rhetoric—direct address, firm assertion—and his avowal of his status as a star on television: "I'm not a doctor, *but I play one on TV*." He is not a medical authority, but establishes credibility by acknowledging this from the outset. At the same time, he invokes medical authority in relation to his fictional role. Through the paradox of truthful avowal transformed into authentication by means of a fictional role, the appeal of the ad is initiated in the unstable mirroring of extratextual and intertextual references: I'm not really a doctor, but I really am an actor; and as an actor in another television text, I really play a doctor.

The impact of the commercial as a consumer message—go out and buy this product—is in part anchored in an understanding of the ad as a moment within television. It refers as much to other shows within the medium as it does to the "real world." This simultaneous referentiality is integral to the comprehensibility of the ad.

5) The viewer is assumed to be familiar with television's modes and genres. An ideal viewer, in this case, will recognize the actor and the role he plays elsewhere on television. Yet it is not necessary to identify the specific show, or role, in which the spokesman acts. It is enough to know that there is an array of television shows that includes doctors as characters, and to surmise that this particular actor (perhaps unfamiliar to me) does indeed play a role on one of these other shows. This premise in turn assumes that viewers know that actors are frequently used in commercials as a basis of celebrity association and appeal.

This detailing of the assumptions and implications of a single advertising campaign can help to clarify the ideological functioning of television. Ideological criticism has its origins in Marxist theories of culture and is concerned with the ways in which cultural artifacts—in
our case television—produce particular knowledges and positions for
viewers. These knowledges and positions link the viewer with, and
allow reception of, the economic and class interests of the television
industry, which is itself part of a broader culture industry (including,
for example, book and magazine publishing, the music industry, and
the film industry). Ideological criticism is based on the assumption that
cultural artifacts—literature, film, television, and so forth—are produced in specific historical contexts by and for specific social groups; it
aims to understand the nature of culture as a form of social expression.
Because of this social and historical specificity, artifacts express and
promote values, beliefs, and ideas that are pertinent to the contexts in
which they are produced, distributed, and received. This may sound
like a simple and straightforward proposition. But in practice, ideological criticism is complicated by a number of factors.

Marxist theory conceives of society as a complex interrelationship
among different practices and institutions. Within society the ways in
which meanings (values, beliefs, ideas) are expressed through cultural texts, and the ways in which these meanings are received and
understood by individuals and groups, is a dynamic process involving
the interaction of multiple influences or determinations. Moreover,
within Marxism, a range of perspectives on culture and ideology has
been developed. The particular approach to ideological criticism that
one uses will vary according to one's position within Marxism. Finally,
television itself is a mass, industrial medium, involving a variety of
texts, produced by many different groups (and individuals), aimed at a
broad and heterogeneous set of audiences. It thus becomes difficult to
talk about a single set of beliefs or ideas that are carried by television in
any simple or immediate sense.

Within Marxist theories of culture and society, the concept of ideology
has been subject to intensive elaboration.[1] Classical (or orthodox)
Marxist theory construes society in terms of a base/superstructure
model. According to this model, the primary and crucial organizing
factor of a human society is its economic *base* (some theorists call this
the *infrastructure*)—its mode of production. Fundamental class al-
liances and material interests are established at this level of society, ac-

cording to who owns, controls, and profits from the basic mode of production. The main class division is between the owners of the means of production and the workers within the mode of production. The dominant mode of production, in turn, determines the *superstructure,* which includes political and legal systems, culture, and ideology (belief systems such as philosophy, religion, or morals). Dominant interests are defined by material interests—the control of economic and productive practices—and are then expressed and manifested in the organization of the superstructure. Because the superstructure is organized according to dominant class interests, it functions to sustain and perpetuate the (current) dominant mode of production.

Thus within orthodox Marxism, ideology refers to ideas, values, and beliefs that may be carried by philosophy, literature, painting, or television programs. The cultural artifacts produced within a given mode of production are seen as reflecting the interests of the dominant class. For example, television—a heavily capitalized and industrialized branch of the entertainment industry—would necessarily reflect the belief system, the ideology, of the dominant class. Viewers are then seen as buying into this belief system, no matter what their positions are within the economic system. From this perspective, a transformation of television's ideology would require a shift in the mode of production—a total reorganization of ownership and control of the medium. The most rigid versions of this approach would not allow the possibility that nondominant views might find expression within commercial broadcast media; nor would they admit the possibility of alternative readings of television programs (which will be discussed later in this chapter, and in the chapter on British cultural studies).

Classical Marxists tend to regard ideology as *false consciousness* or mere illusion. The ruling class promulgates systems of ideas and beliefs to promote its own interests that are mistakenly adopted by oppressed or subservient classes as their own. In adopting the values and beliefs of the ruling classes (those whose interests are served by the current system), the subservient classes participate in their own oppression. "Real" ideas about society and class values, the argument goes, can be achieved only through a materialist analysis of the economic base—an analysis to reveal the actual class dynamics at stake in a given institution or system. Political activism and social transformation can occur when ideology is exposed as such, and the truth of a materialist analysis is brought to light.

A rather simple version of classical Marxism, applied to television, might argue that although the working classes believe that television is harmless entertainment, offering a pleasant way to relax at the end of a hard work day, in actuality it lulls them into passive inaction and indeed instills bourgeois values. Thus these working class viewers exist in a state of false consciousness. By failing to recognize how their ideas and values are formed for them to serve the interests of others, they are dupes of ideology. Corrective political action would involve educating the working classes to understand how the medium carries values at odds with their class interests and that their potential to act as a class is blunted when they watch television to "relax."

Within Marxist thinking, the theory of ideology as false consciousness has been subject to criticism and revision. On the one hand, it does not explain how or why people so readily adopt ideas that they know are at odds with their own interests in society. At the same time, it is clear that the ruling class does not exercise force to *narrowly* restrict the expression of beliefs and ideas through media. In other words, classical Marxism does not provide an adequate theory of social subjectivity to explain how individuals assume a position in society. Moreover, by emphasizing institutional/economic analysis of media and concentrating on the expression of overtly political ideas in media, the classical Marxist approach largely ignores the fact that most people watch television, most of the time, because they find it enjoyable. In this sense classical Marxism has not developed critical or theoretical perspectives for dealing with the pleasures of watching TV. Because of these problems and limitations, many theorists have acknowledged the inadequacy of a definition of ideology as merely false consciousness and have developed alternative perspectives on ideology. These alternative approaches variously stress contradictions within society, the coexistence of competing ideological positions, and the ways in which individuals assume positions in relation to their social world. In the process they offer a basis for an ideological approach to understanding television and its pleasures that is not reducible or limited to false consciousness.

Still working within the base/superstructure model, some theorists have emphasized the principle of *uneven development* (present in Marxist thought from its inception), recognizing that social transformation is a constant but inconsistent process. All parts of the social system—the mode of production and superstructural organizations—are dominated by ruling-class interests, but traces of earlier social forms, as well as elements of more progressive forces, coexist alongside the dominant. Moreover, the contradictory and conflicting perspectives are unevenly distributed.

The emphasis on uneven development allows for a more complex understanding of society and ideology. The legal system and scientific thought are both part of the superstructure, and may primarily embody or express dominant class interests. But they do so in different ways. This perspective also foregrounds the fact that a variety of voices may express conflicting class interests, although the ruling class interests will prevail in most contexts. Italian Marxist Antonio Gramsci used the term *hegemony* to explain the complex ways in which the dominant class maintains control over society. Hegemony describes the general predominance of particular class, political, and ideological interests. Although a society is made up of conflicting class interests, the ruling class exercises hegemony in that their interests are accepted as the prevailing ones. Social and cultural conflict is expressed as a struggle for hegemony, a struggle over which ideas will be recognized as the prevailing, common-sense view.

A more thoroughgoing reformulation of ideology was developed by the French Marxist philosopher Louis Althusser, who reconceptualized society through a revision of the base/superstructure model itself. The perspectives on television that I develop in this chapter, although they draw on Marx and Gramsci, are based on Althusser's work. The following summary of the Althusserian position is somewhat schematic, but it provides a context for understanding ideology as a social practice. As a Marxist, Althusser recognizes the importance of the mode of production in determining the nature of society. But rather than arguing that the mode of production is a *base* that unilaterally determines the rest of the superstructure, he argues that society is comprised of a variety of interrelated social and intellectual activities or practices, including the economic, the political, and the ideological.[2] Together these different practices comprise the *social formation.*

Economic practice involves the mode of production—the nature of productive forces and the relations of production. Political practice describes social relations and forms of social organization. Economic and political analyses are therefore concerned with the nature and relations of power expressed in particular economic and social systems. Ideological practice refers to systems of representation (images, myths, ideas), in which individuals experience their relation to their material (economic and political) world.[3] Ideological analysis aims at an understanding of the ways in which meanings are produced by and for individuals within a social formation. Economic, political, and ideological practice are distinct but coexisting arenas of human activity. They exert mutual influence and pressure on one another, but also operate with *relative autonomy.*

The idea of relative autonomy is the crucial revision Althusser introduces in relation to the orthodox Marxist base/superstructure conception of society. Although economic practice ultimately determines all other practices (or, as Althusser puts it, the economic determines all other social practices "in the last instance"), political and ideological practice do not function as direct expressions of the class interests defined by economic practice, but have a life of their own. That is, each sphere of social practice has its own structures, dynamics, and history and must be understood in its own terms. Because each sphere of social practice has this relative autonomy, political and ideological practice are important arenas for contestation among social interests and groups along with economic practice. The transformation of society requires active intervention in all areas of social practice and cannot be limited to the domain of economic relations.

Moreover, all of the social formation—its various practices and their relations—is characterized by disunity or contradiction. That is, social practices must be understood as complex and heterogeneous structures rather than as simple and unified entities. For example, above I refer to the idea that each arena of social practice is a site of contestation among social groups. But we must recognize that the very idea of social groups is a complex construction. We cannot assume that one set of terms explains social identity. On the contrary, there are many different ways of understanding groups with conflicting, intersecting, and parallel interests. One crucial term of distinction—class—is established as a function of economic practice; but other terms may emerge specifically within the contexts of political or ideological practice. A given individual may be defined and positioned by a variety of categories—including class, profession, nationality, age, gender, race, and so forth. At times the various interests of an individual, defined as an effect of these intersecting categories, may work in concert, whereas at other times they may be divided or come into conflict with one another.

The Althusserian conception of society and ideology is not without problems and limitations and has itself been subject to criticism and revision.[4] However, for our purposes, the basic terms of his understanding of ideology have important implications. Because Althusser defines ideology in terms of both systems of representation and individuals' relations to their material world, his theories have been useful and influential in film, media, and cultural studies. The relative autonomy of ideological practice in this context points to the importance of studying modes of representation and recognizing that they are socially determined, but are not simple or direct reflections of dominant class interests understood strictly in economic terms. Furthermore, because ideological practice concerns the relation between individuals and their social formation, it focuses attention on individuals as social subjects who both construct and are constructed by systems of representation.

Indeed, insofar as ideological practice designates the ways in which individuals experience meaning—how we produce meanings about ourselves, for ourselves, in representation—there is no such thing as being "outside" ideology. Or, as cultural theorist Stuart Hall has said, "The notion that our heads are full of false ideas which can, however, be totally dispersed when we throw ourselves open to 'the real' as a moment of absolute authentication, is probably the most ideological conception of all."[5] What Hall points to here is the way in which ideology presents itself as "natural" or serves to naturalize a given system of representation. "When we contrast ideology to experience, or illusion to authentic truth, we are failing to recognize that there is no way of experiencing the 'real relations' of a particular society outside of its cultural and ideological categories."[6] In other words, the point of ideological criticism is not to find unadulterated truth or unbridled manipulation "beneath" or "behind" a given text or system of representation, but to understand how a particular system of representation offers *us* a way of knowing or experiencing the world.

For example, the shot/reverse shot construction in film and television is often discussed in these terms and can help clarify how a system of representation gives individuals a particular perspective on events—a perspective that *seems* natural. This pattern of cutting back and forth between two characters in close-up (or medium shot) is frequently used for dialogue sequences and is considered the "natural" way to represent a conversation. However, it is in fact a highly artificial construction that in no way duplicates any one's experience of "real" space. As the image cuts between two characters, the visual space is fragmented and the angle of vision offered to the viewer is not the perspective a character in the scene would share. In this sense it is a perspective created for the viewer outside the fiction. Thus, although we are so accustomed to seeing dialogue scenes rendered in shot/reverse shot fashion that we accept it as "natural" or "realistic," it is realistic only in relation to the conventional system of representation in which it is used. In this context, a dialogue shot in a static long take seems "boring" and "artificial" by contrast to the more familiar shot/reverse shot construction, because we are used to being "naturally" directed where to look, and when, in shot/reverse shot scenes (though the long take is simply another, less familiar, perspective constructed for us).

Given that ideology involves a complex set of practices and relations, ideological criticism includes a variety of procedures and methods that may emphasize different aspects of the individuals/systems of representation/social formation network. A mass art form such as television provides a crucial and fruitful arena for ideological analysis precisely because it represents the intersection of economic-industrial interests, a system of texts, and a leisure-entertainment activity. Marxist scholarship in mass communication, especially before 1980, overwhelmingly centered on economic and institutional analysis of media systems—the result of scholars working within the orthodox Marxist tradition. However, there is a growing interest in a text-oriented ideological criticism that examines television as an ideological practice—that is, as a complex system of representation through which individuals experience and understand their world. With this emphasis, critics have drawn on a range of analytic methods—including economic, textual, and cultural approaches—in an effort to clarify how and why the medium carries the meanings and values it does for its audience. This more recent ideological criticism is concerned with the ways in which a particular text or group of texts functions as a part of ideological practice and offers a system of knowledge or a way of experiencing the world for a viewer. Ultimately its goal is to understand how textual systems, with their relative autonomy and disunity, also function within the dynamics of a given social formation.

THE VIEWER AS CONSUMER AND AS COMMODITY

Even a cursory glance at American television—in terms of both its textual system and its economic supports—reveals that advertising holds a central position. Networks and individual stations earn profits by selling time to commercial sponsors. In the United States, television followed the model of the radio industry in developing networks and commercial sponsorship, and from the start considered the viewer/consumer on a national scale. Although consideration of the development of the television industry and its industrial-economic infrastructure is beyond the scope of this chapter, the position and functioning of advertising is a crucial aspect of ideological analysis, especially as it is a manifest part of television programming.

With the prominent and regular display of commercials on television, the source of network and station income is not hidden but is, on the contrary, an integral part of television program flow. The importance of commercial sponsorship is underscored in popular television magazines and newspaper columns that report on ratings and explain (regularly, if intermittently) the intense competition for viewership in relation to network and station advertising revenues. Although the particular forms advertising has assumed in commercial television are by no means natural, but the result of specific historical and institutional developments, these forms are nonetheless a "fact" of contemporary television viewing. American commercial television is "free"; viewers do not pay for broadcasting through a license fee (as is the case in Britain) because advertisers pay for time to promote their products. Because commercial television is first and foremost a mass-advertising medium, the viewer is positioned as a potential consumer—one who will presumably, or hopefully, purchase the products promoted on television. (Of course, advertising expenses are calculated in the wholesale pricing of products. In this sense television is indirectly supported by all consumers, whether or not they watch television or see commercials for the products they purchase.)

To be addressed as a consumer does not mean that every viewer is in the market for, or will purchase, everything—or anything—that is advertised on television. Rather, the address to the viewer as a consumer means that s/he is regularly subjected to a range of appeals for a variety of products. Even viewer-supported cable services such as Home

Box Office, Showtime, and so forth, include promotional spots for their own programs. In these instances, the viewer is addressed as a potential "consumer" for the station itself and its future programs, with the aim of encouraging the viewers to continue watching and/or paying for the particular service.

The address to the viewer as a consumer in general can be channeled in a variety of ways. For example, certain programs or programming periods are seen as having particular demographic audiences, and a large proportion of the advertising is designed to appeal specifically to them. Commercials broadcast during Saturday morning "children's" programming are decidedly different from those that might be aired during a soap opera, the evening news, or a golf tournament. Similarly, the viewer is often addressed as someone who watches and has watched television regularly, and is therefore assumed to be familiar with an array of stars, characters, programs, and genres on television, including other commercial spots. The cough medicine commercial that served as our initial example invokes the knowledge of television's fiction programs. A recent series of ads for Tostitos corn chips intercuts footage from various "old" television programs such as *Leave It To Beaver* and *Mr. Ed* with new footage to create the illusion that the product's spokesman is talking to characters from these shows. Many commercials for different brands of the same kind of product or service actually parody one another. Brand name differentiation is thus linked to a specific textual strategy, norm and parody, in which it is necessary to know the norm in order to recognize the parody. For example, MCI—a long-distance phone service company—has designed several commercials as parodies of AT&T ads.

In all of these cases, television viewing and consumption become so strongly affiliated that it is difficult, if not impossible, to differentiate between watching television for the purposes of entertainment, information, or relaxation, and recognizing one's place within a consumer society. From this perspective, analysis of specific ads or advertising campaigns in terms of their references to actors, characters, and programs from the medium in general can be a useful way of initiating ideological analysis of the medium. This approach can be further developed in the study of tie-ins between television programs and commercial products, and the ways in which whole product lines are developed in conjunction with programs—products that are in turn advertised on television (for example, Smurfs and Forever Krystle cologne/*Dynasty*).

Ideological critics are not simply concerned with the ways in which television addresses its viewers as consumers but push the implications of this process further to describe how viewers are, in turn, transformed into commodities by the television industry. An elaborate apparatus of ratings is in place—the Nielsen and Arbitron ratings systems are the most prominent—to measure the audiences for specific programs and stations. Ratings are crucial to the television industry, which uses the statistics to determine network and local advertising rates. The viewer-as-consumer is thus abstracted into an object of exchange value that the network or station offers to a commercial sponsor. We get "sold" to advertisers in lots of one thousand. Individually, we might figure merely as part of the "mass" audience or as part of a particular demographic group with identified purchasing power.[7] Networks and stations do not simply sell blocks of time, but times for specific programs with specific projected audiences. In these terms the individuality of the viewer, and the heterogeneity of an audience, are secondary to the exchange value potential of so many thousand, or million, viewers.

Understanding the television viewer as, at once, a consumer and a commodity provides a basis for analysis that draws together the culture industry on the one hand, and consumer society on the other. One does not normally decide to watch television in order to look at possible products for purchase, or to become a token in the system of exchange between networks, stations, ad agencies, and commercial sponsors. Yet it is inevitable that both of these positions are at stake to support and sustain an activity that is undertaken for a variety of other more conscious reasons—to relax, to see a particular sports event, to learn of the day's events, or because there is nothing better to do. From this perspective, ideological analysis emphasizes the commercial message as the linchpin between television as information/entertainment and television as an industry, with the viewer as the site where these meanings, or forces, converge. Yet while an awareness of how the material interests of the industry are most directly expressed on television leads to an understanding of the viewer as a consumer and a commodity, this does not exhaust or complete the work of ideological analysis. On the contrary, it becomes the grounds for raising a range of issues and questions focused on texts and readers.

However abstract or impersonal the implications of commodification may be, viewers are not forced to watch television, but choose to do so "freely," as individuals. This choice takes place by and large with some awareness of the process by which one becomes a consumer/commodity in the very act of viewing. In light of this, ideological critics have turned their attention to questions about the nature of the meanings and the pleasures television offers through its programs. For in the absence of force, one may assume that this is how the medium attracts its audience in the first place—that it offers a certain familiarity, comfort, and diversion to which many individual viewers can respond. And this attraction can occur even with the relative self-consciousness that one is quite literally being sold a bill of goods in the process of watching television.[8]

IDEOLOGY IN NARRATIVE

If the viewer is not coerced into watching television, the question of why people watch—the nature of the medium's pleasures—becomes a crucial arena of investigation for ideological critics. It is common to deride the medium and its programs as trivial, simple-minded, crass commercialism aimed at the "lowest common denominator" of intelligence. Despite being told they are "couch potatoes" by media critics, people still watch television in large numbers and with great frequency. The medium's convenience and accessibility furnish a partial explanation for its popularity, as television programs are quite literally at one's fingertips. But if this is a necessary precondition of television's effectiveness as an agency of consumerism, it is not sufficient to account for the values and meanings the medium may hold for its viewers or for the pleasures viewers might derive. For these we must turn our attention to the programs themselves and see what they have to offer, individually and as a group.

The analysis of individual programs, groups of programs, and viewer-text relations is central to any understanding of the appeal television holds for its audience. Here ideological criticism draws on the methods and insights of different approaches to textual analysis—semiotics, genre study, narrative analysis, psychoanalysis, and others—to understand what meanings are made available through the medium and its programs and the nature of viewer engagement. In drawing on these various methods of analyzing texts, the ideological perspective assumes that television offers a particular construction of the world rather than universal, abstract "truth" of idiosyncratic individuality. In other words, ideological criticism examines texts and viewer-text relations to clarify how the meanings and pleasures generated by television express specific social, material, and class interests. This is not to say that a given program or episode directly expresses the beliefs of a particular producer, writer, director, or network programmer (though obviously these may be contributing influences and viewpoints), nor that there is some conspiracy among television executives to control the ideas expressed through the medium. Rather, an ideological perspective focuses on the systematic meanings and contradictions embodied in textual practices—for example, the way familiar narrative, visual, or generic structures orient our understanding of what we see and "naturalize" the events and stories on television.

Narrative and generic conventions are crucial ways in which television handles social tensions and contradictions. Thus, ideological criticism might begin with a narrative analysis to see how the structural and functional logic of plot development in an episode is used to explain and naturalize a sequence of events. This analysis may reveal values or meanings that are not immediately apparent in, but are crucial to, the general coherence and logic of the storyline. A discussion of a specific episode of *Webster* may clarify this point. The program is a family sitcom centered on a young black boy, Webster Long, and his white foster parents, George and Katherine Papadopoulous. (The premise of the show is that Webster's parents were killed in a car accident while George and Katherine were on their honeymoon. In the premiere episode they returned home to discover Webster in their custody. George is a sports newscaster, a former pro football player who was a best friend and teammate of Webster's father. Katherine is an upper-middle-class woman who works in the city government.) In the episode broadcast 17 January 1986, a neighbor's habit of playing the state lottery excites Webster's attention. He is convinced that if he can buy a ticket he will become a millionaire. George tries to convince him that gambling is a waste of money. But in the face of Webster's persistent enthusiasm George allows him to spend his allowance on a lottery ticket. Webster chooses his six numbers with his family, as they each pick their lucky number and their age. Katherine goes last, and instead of revealing her age she volunteers to buy the ticket and fill in this last number herself.

The night before the drawing Webster dreams that he has won the lottery. His dramatized fantasy is an exaggerated, parody version of what it means to be rich. Servants lead him around his mansion on a horse for amusement and do his homework for him. Webster sits amid ornate antiques in a red silk robe trimmed with gold sequins—accentuated by the use of star filters—and offers lavish presents to his parents and their friends, including an immense pearl that he gives to Katherine. (He explains that it is left over from the pearl necklace he had made for the Statue of Liberty.) It is too large to wear, but the perfect size for bowling in the mansion's indoor alley. Towards the end of the dream, as Webster revels in his wealth—noting that the government has put his face on a new, trillion-dollar bill—Katherine reminds him that "When you give out of love, you're rich even without money." The next day the whole family watches the lottery drawing on television. As the numbers are called one by one they directly follow Webster's ticket, ending with Katherine's age, thirty-nine. George begins to celebrate until Katherine reads the ticket she purchased, where the sixth number is thirty-six. She confesses that she lied about her age when she finished filling in the ticket, so that Webster does *not* win millions of dollars. In the final scene the family is commiserating with one another over their loss. Webster seeks to console George and Katherine, and Katherine notes that it doesn't matter if they have money. "If you've got what we've got, you can be rich without money." Webster repeats her statement, and explains that Katherine said the same thing in his dream. "We could have all the money in the world and not be as rich as we are," he affirms.

On the one hand, we are presented with a moral tale about the value of gambling, even in legal forms. George insists that no one ever gets rich through games of chance, that playing the lottery is a waste of money, which in fact proves to be the case. Indeed this is the meaning of the episode as summarized in the weekly *TV Guide* listing for the show: "To teach Webster how hard it is to get rich playing the lottery, George buys him a ticket." (One might note a certain slippage of summary here—in the episode George actually agrees to let Webster spend his allowance money on a ticket, which Katherine purchases for him.) On the other hand, this linear and predictable development is cut across and displaced by another that promotes Webster as a privileged, almost magical agent (a rather common attribute of children on television), as his intended scheme for picking numbers proves to be effective: the family members' lucky numbers and ages were the winning numbers for the week. His childlike faith in his ability to win is thus confirmed by the narrative outcome of the lottery drawing, strong enough to transcend the "adult" message about the serendipity of gambling.

Indeed Webster would have won millions if Katherine had not betrayed his scheme by misrepresenting her age, which she is able to do, in the guise of helping Webster achieve his goal, by actually purchasing the ticket. This particular narrative move relies on a cultural stereotype—women lie about their age—to naturalize an outcome that sustains the double logic indicated above; gambling is shown to be a waste of money and yet Webster maintains his privileged status in surmounting the odds in principle, if not in fact. George and Webster can *both* be proved "right" by the narrative because of Katherine's toying with the formula for picking numbers, which she does even though, within the fiction, only three characters would even know that "39" on the lottery ticket referred to her age. To reinforce the "naturalness" of her lie, she offers the following explanation when apologizing: "I don't know what got into me. It was like a reflex." An additional implied aspect of this sequence of events is that Webster's magical faith can only work *once,* in the context of his initial naive belief in his ability to win. From now on George's perspective on gambling will prevail. In the process Webster shifts his interest in wealth as money to an emotional investment in wealth as familial love.

One might further develop an ideological understanding of the episode specifically in terms of the program's construal of "wealth" in this particular story. On a week-to-week basis, the general lifestyle of the family—their house, clothing, occupations—represents a recognizable upper-middle-class image. In this episode, the dream of becoming an instant millionaire is first carried only by Webster, but is adopted by George and Katherine in the course of the lottery drawing. As the numbers are drawn they all get increasingly excited, so that they are all profoundly depressed when they realize that they "lost" by one number. In this way, the episode implies that the style of living it regularly represents is simply normal, and that the lure of millions of dollars offered by the lottery is a fantasy shared by everyone—all families, all conceivable viewers—in the same way.

Within this context of upwardly mobile class aspirations, Webster's dream is obviously parodic, a conglomeration of *media* representations of the very rich—lots of servants, a live horse—with childlike additions.[9] The pearl, above all else, condenses the admixture of imagery insofar as it represents a precious gem (fetish object of congealed wealth), jewelry (but not usable as such), a sport (to link up with George's profession), and a toy all at the same time: no real pearl could ever be that size. As such, Webster's fantasy of riches is something that no one—or perhaps only a child—would really want, or would only want as a fantasy. The absurdity of Webster's fantasy helps to soften the blow of not winning the state lottery, as does the repeated dictum that love (implicitly familial love) itself constitutes wealth, rather than money. This homily would seem to be the "message" of the episode, especially when it is reiterated by Webster as the final agency of authority, as he shifts his privileged, magical faith in the lottery to the family.

But a more detailed analysis of the episode's narrative logic indicates that it is only one stage or moment in a more intricate scheme of values and meanings. The idea that familial love may be expressed in *honesty* (versus lying) to promote the realization of fantasies is implicitly an issue developed in the episode's parallel subplot. At the opening of the episode, George comes home depressed because his favorite Greek restaurant has been closed, and he can no longer spend Friday afternoons eating his favorite Greek dish prepared by his "Yaya" (Greek grandmother figure). Katherine traces his Yaya so she can learn how to make the dish for George. The Yaya comes to their home while George is out and suggests that Katherine let her do the cooking; then Katherine can pretend she made it herself. But Katherine refuses to go along with this idea (even though, throughout the program, she has been defined as incompetent in the kitchen), insisting that she could not lie to George about something like this. In this case, honesty wins out, as Katherine successfully prepares George's favorite dish, and is thereby able to restore his Friday afternoon ritual. This plot development is embedded in, and secondary to, the lottery story, in which Katherine lies about her age, and thereby fails Webster.

The theme of familial love thus supports or frames the overall logic of the episode's narrative development, but it hardly begins to explain or contain the network of ideological values constructed through this particular ordering of events. Rather, our analysis has suggested the importance of recognizing a combination of narrative functions as the work of ideology. Some of these are specific to this particular show—for example, sustaining the privileged status of the character for whom the program is named. Others have more to do with general or typical practices of representation within the medium, such as implying that an upper-middle-class, elite lifestyle is "average." At other moments the show invokes broader cultural or social myths, in this case that women do not like to admit their age, especially as they approach forty. All of these are drawn together and activated in this episode to naturalize and give sense to a story with a more overt moral message about gambling, wealth, and the family. Their interaction and particular configuration account for the ideological meanings of the episode. Thus, ideological criticism aims at an explanation of the narrative, visual, and generic strategies that support and sustain the overt values and messages that may emerge at first glance.

This approach is not limited to dramatic narrative programs but is equally pertinent to game shows, news, documentary, sports, and other kinds of television programming. Game shows, for example, offer structured arenas of competition, most often with the goal of winning lavish prizes. As such, they promote and extend the consumerist basis of the medium in varying degrees. A program such as *The Price Is Right* directly involves consumer knowledge as the basis of competition, with success or failure based on contestants' ability to estimate the retail value of a wide range of products including furniture, cars, vacations, jewelry, groceries, and household appliances. In the course of proceeding to the grand prize—the showcase showdown—participants are subjected to a variety of games that require them to demonstrate their skill as a consumer. In a crucial sense, the whole show becomes a sort of continuous advertisement as each new object and product within these games is described by brand name and qualities, often with a promotional tag line.

While money and prizes remain the goal of most game shows, they do not all so blatantly exhibit consumerism as the terms of competition. Other game shows structure knowledge within a restricted field. In shows such as *Password Plus, Wheel of Fortune,* and *The $100,000 Pyramid,* the ability to guess the right word, phrase, or category on the basis of the least information defines the structure of competition: How many letters, words, or definitions are required before you can properly identify the correct answer? These structures are charged with significance in the context of game show competition, as players strive to fill in the blanks first to reap the rewards of winning. At the same time, most game shows integrate elements of chance in the course of play; within the terms defined by the show, skill is necessary but rarely adequate for achieving success. Thus, for example, contestants on *The Price Is Right* are called from the studio audience. Whether or not you even get to compete is a matter of luck. In other game shows, the amount of money to be won may be determined by the spin of a wheel or by pressing a button at the right time. Chance is even incorporated into the game shows that emphasize knowledge or skill, to the extent that contestants usually have to pick categories blindly, without full knowledge of the kinds of information that will be required. On *Wheel of Fortune*, contestants not only compete to correctly identify the person, place, thing, or phrase contained on the game board in a variation of hangman, but spin a wheel every time they request a new letter to plug into the empty spaces on the board. The wheel determines the amount of money each letter will earn but also includes slots for "Lose a Turn" and "Bankrupt"; a player proceeding with all due skill can suddenly be left out of a crucial round of the game. In other words, a double narrative logic is at work—a combination of "knowledge" and "luck"—which is typical of most game shows, though the balance between elements of chance and skill may vary.

This dual logic produces the possibility for evoking familiar adages or versions thereof: life isn't fair; success is a question of being in the right place at the right time; it's not what you know; and so on. At the same time, one can admire skillful players, compare contestants in terms of how well they play the game, and even measure oneself against them. But this occurs against a backdrop of acknowledged serendipity. One can aspire to the prizes and simultaneously console oneself: "I may be a better/worse player, but I might have better/worse luck if I were actually competing." In other words, in most game shows neither sheer skill nor sheer luck prevails, and this contributes to their effectiveness and appeal. We can enjoy the adept players without feeling hopelessly stupid, because we recognize that luck has something to do with their success, and we can maintain feelings of superiority over lesser players whether or not the game's elements of chance work in their favor. This sustains our pleasure as we watch a particular show, mentally participate in the play, or root for a particular contestant.

Thus an ideological approach to game shows, like that of dramatic programs, aims at an understanding of the underlying narrative logic and patterns that structure the games and hold our interest. It is not only a question of consumer rewards—big money, new cars, living-room furniture—but also of the ways in which these are achieved in a regulated field of competition, and the nature of the rules of the game. What kind of "knowledge" is at stake in the show? What sort of competition is involved? How are these incorporated and intertwined in a series of steps en route to the grand prize? These are the questions one must ask in order to initiate an understanding of the ideological meanings produced in this genre.

IDEOLOGY AND CONTRADICTION IN THE TEXT OF TELEVISION

The discussion of *Webster* and of game shows indicates that ideological analysis is not necessarily a simple or self-evident practice, even when various levels or moments of ideological meaning work more or less in concert in the course of a given program. Similar approaches to other episodes and programs may reveal the production of ideological meanings that are less stable, less unified. An understanding of more diffused ideological meanings can in part be seen as integral to the structure of the medium, which usually aims to attract the largest possible audience for any given program. From this perspective, textual strategies that allow a limited plurality of interpretive perspectives enable a broader basis of appeal.

At the same time, the production of multiple ideological positions can be seen in terms of programming practices, as individual episodes and programs are situated within program flow. The texts that comprise television are not discrete and delimited but are juxtaposed with and bump up against one another. Individual episodes are segmented and interspersed with commercials, news briefs, and program previews, which are themselves sequences of mininarratives. All of this is, in turn, positioned within an unceasing flow from program to program. This appearance of an endless text is regulated through various kinds of repetition—the same shows in the same position within the weekly schedule, genres, and so forth. In this context, a given program may develop variable perspectives and issues over time (a process that is especially evident when the show is presented in syndicated reruns on a daily basis). Thus, within a single episode, evening, or season of television, the ideology of particular programs may emerge as variable, slippery, or even contradictory.

Cagney and Lacey offers striking examples in this regard, but is hardly singular in offering heterogeneous ideological meanings to its viewers. The program features two female police detectives as the center of narrative interest and espouses a sympathetic liberal feminism. Individual episodes frequently foreground personal and professional issues that are perceived as being of particular concern to women—sexual harassment, problems of working mothers, child abuse, and so forth. Yet the visual and narrative strategies engaged in individual episodes may work to undercut or contradict the ostensible progressive orientation of the show, relying on conventional modes of plot structure and visual representation that have been seen as undermining the power and effectiveness of women. For example, framing and mise-en-scène are sometimes used in ways that imply that one or the other of the central characters is caged or trapped. This may produce an impression of weakness or helplessness on their part, even though as narrative characters they are supposed to be active, competent detectives.

In one episode, Cagney initiated a sex discrimination suit against a superior officer, against the wishes and powers of the New York City police department. When she resolved to pursue the case, she was seen in a close shot, framed against the barred windows of the precinct interrogation room. In another episode, Cagney was physically threatened by a suspect in a murder to which she was witness. He followed her around in an effort to persuade her, through sheer threat of force, not to testify against him. This episode included repeated shots of Cagney isolated in her apartment, almost cowering, trapped by the camera and the suspect as he watched her through the rooftop skylight. In this instance, her ability to perform as a cop (aggressive, strong, confident) was displaced by conventions for representing women as subject to the menacing threat of a narrative character and the look of the camera. Thus, even as the program offers two "strong, professional" women, it deploys familiar visual and narrative conventions—an established visual and narrative language—that restrict women's ability to control their own fates and subjugate them to the control of the camera and the male gaze.

These examples forcefully underscore how television can be analyzed in terms of disunity and contradiction, as the codes of narrative and visual construction come into conflict at particular moments within the program. Ideological criticism aims at an understanding of these contradictions as constitutive of the text's ideological problematic—the field of representational possibilities the text offers.[10] As

David Morley explains, "The problematic is importantly defined in the negative—as those questions or issues which cannot (easily) be put within a particular problematic—and in the positive as that set of questions or issues which constitute the dominant or preferred 'themes' of a programme."[11]

In the case of *Cagney and Lacey,* the problematic is initially established as a function of a number of generic and discursive systems. On the one hand, the program combines the police drama with aspects of domestic melodrama. As a police show, the focus on urban crime, police procedures, and so forth provides a context for a certain range of issues: the ethics of dealing with informants, the role of the press in reporting crime, and the impact of crime on its victims, for example. On the other hand, the focus on two women detectives—one married with children, the other single, and both nearing forty—accounts for the examination of domestic and interpersonal issues. With this emphasis, the problems involved in raising children or of balancing careers and families can be raised. Simultaneously a "feminist" discourse cuts across both of these, and at times offers an explicit connection between them, as the show quite consciously addresses issues of concern to women in a "progressive" spirit. For example, the show may include concerns of pornography or child abuse as part of its police plots.

Together these areas begin to define the ideological problematic of *Cagney and Lacey*—the kinds of plots it will include and the nature of the issues it will raise on a week-to-week basis. The particular combination establishes its similarities to other programs within television but also differentiates it from other police shows on the one hand (which may not share its domestic and feminist concerns), and from shows that share its domestic or feminist concerns but are not police shows on the other hand (for example, *Family* or *Kate and Allie*). Within its defined ideological problematic, as specific topics or questions are raised, the program may orchestrate a variety of perspectives, without clearly insisting that only one position is acceptable. For example, in one episode Cagney thinks that she might be pregnant. She has to consider an array of options—whether or not to tell the potential father, whether or not to try and marry, whether or not to have the baby—and decide on a course of action. These possibilities and their implications are explicitly raised, often in discussions with Lacey, while pursuing the police plot. However, it turns out that she is not pregnant after all. Thus the choices prove to be hypothetical options (not requiring a decisive course of action) rehearsed by the program, a way of addressing issues of concern to "modern women" (and combining the program's feminist and domestic voices) in the abstract.

At the same time, in the elaboration of a particular problematic, the field of choice is circumscribed; although different perspectives may be introduced, they are not infinite. For example, in *Cagney and Lacey* feminism is explored in the context of the traditional, middle-class, nuclear family. Although Chris Cagney has no husband or children, her familial situation is explored in relation to her father and brother (the former a semiregular character, the latter featured in a number of episodes). Similarly, the police system itself and its hierarchy of authority may at times cause problems for the protagonists, not only in the episodes dealing the Cagney's sexual harassment case, but also in conflicts over areas of jurisdiction between divisions or precincts. This strategy allows the program to raise questions about the police system, or to suggest that it has problems, but it never poses a thoroughgoing challenge to the system itself. Instead, dramatized problems are seen as weaknesses or aberrations in a fundamentally good (natural and necessary) system, within which the characters of Cagney and Lacey represent an ideal.

The latitude of competing voices and positions within the established problematic presents itself as a "totality" because different points of view are incorporated, even though it is a delimited or circumscribed range of choices to begin with. Moreover, these multiple positions and points of view are often regulated by an implicit hierarchy that privileges certain positions over others. This can be seen as a strategy of containment, as minority positions or deviations from the mainstream are introduced but are framed and held in place by more familiar, conventional representations. John Fiske and John Hartley describe television's cultural mediation as working to *claw back* any subject to a central focus by referring the subject at hand to familiar modes of understanding. "This inevitably means that some features of the subject are emphasized rather than others. For example, nature programmes will often stress the 'like us-ness' of the animals filmed, finding in their behavior metaphoric equivalences with our own culture's way of organizing its affairs."[12] In relation to *Cagney and Lacey,* the roles held by the program's title characters are hardly aberrant, but they are not typical within the context of television—giving women an unusual degree of strength and independence in terms of their narrative roles that might be seen as a challenge or threat to traditional gender roles. The use of conventional visual strategies for representing women, along with the domestic plots that emphasize the more traditional roles of wife, mother, and daughter, can be seen as working to contain this potential threat.

The movement between program plot segments and commercial breaks may exacerbate the sense of contradiction, especially when the ads employ typical depictions of women and include a substantial number of appeals for products aimed at female consumers (cosmetics, stockings, household cleaning products, etc.). In an episode broadcast during the 1983–84 season, the professional plot concerned the illegal adoption market. The parallel personal story focused on Detective Lacey and her family, as she arranged for temporary custody of the abandoned infant whose plight prompted the investigation. The wealthy couple who had "purchased" the child left her at a hospital when they discovered she was deaf. In the course of the investigation, the baby's real mother was located, and in a confrontation with Lacey she explained that she had sold the baby out of economic hardship, but now regretted her actions and hoped to reclaim custody. The episode concluded with Lacey returning the infant to a child welfare officer in an extremely emotional scene, on the heels of the Lacey's decision to look into the possibility of adopting her.

The commercial that directly followed this scene was for Hallmark Mother's Day cards, featuring an infant and a jingle about a first Mother's Day. The highly sentimentalized Hallmark version of motherhood was jarring in the face of the program's portrayal of motherhood. Indeed, the episode offered a multiply problematized representation thereof—a wealthy woman who can't bear children, a welfare mother forced to sell her child out of economic necessity, and the professional mother who would like another child and invests in that possibility only to experience a loss. (It is interesting in this respect that during the 1985–86 season Lacey gave birth to a third child, a daughter—a narrative development in part determined by actress Tyne Daly's pregnancy. Thus the problematic of motherhood, part of the domestic emphasis of the show, reemerges over time.) The discrepancies in conception between the program and the commercial seem irreconcilable; but they are mutually interdependent, as the troubled versions of motherhood represented in the episode appear to be progressive or realistic, in contrast to the more traditional, sentimental representation offered by the Hallmark ad. Moreover, these representations work together within the problematic of "motherhood," with the ad serving to provide a necessary supplement in the overall concert of voices within this rubric.

Contradictions between juxtaposed segments of television flow are not necessarily systematic in the sense of being willfully or consciously planned by programmers or sponsors (though on occasion one suspects conscious planning). However, they occur regularly, if individually, throughout the course of television programming. Almost everyone can cite particularly striking examples. A network public service message aired on Saturday morning about health and nutrition might be embedded in a series of ads for candy, cookies, or sweetened cereals; or a news story about new research linking smoking with some illness might be followed by an ad for smoker's tooth polish.

At the same time, with regard to a program such as *Cagney and Lacey,* some feminists have stressed the importance of "female bonding" represented in the relationship between the two main characters as an important aspect of the show. Although the two characters bring different and often conflicting perspectives to bear on issues of personal and professional life within the fiction, they work successfully as a team and provide one another with mutual support. They do not compete against one another, but negotiate and combine viewpoints in order to work together. The narrative occasions that foreground this sort of interaction are seen as privileging women's perspectives and as offering the possibility of concerted action grounded in different aspects of women's experience. In these instances it is possible to argue that the program goes beyond "strong" images of women, offering a nascent feminist ideology within the context of mass art.

Thus, over time, within and across individual episodes, *Cagney and Lacey* produces a range of ideological effects and meanings. The contradictions and multiplicity of views help explain its appeal to a broad potential audience because one can recognize progressive, liberal, and traditional values working at once through the fabric of the show. Depending on where and how one focuses one's attention, a range of belief systems can be partially satisfied and fulfilled, though they are received in a context of contestation, moderated by the other perspectives that accompany them. An awareness of this field of multiple meanings as the work of ideology is crucial in understanding the effectiveness and appeal of television as a mass medium.

On an even broader scale, it is possible to look at a whole series over time and see how certain fixed ideological structures recur or sustain unity over the variations and individual stories of single episodes. *One Day At A Time,* for example, is no longer in current production, but it is available for viewing in syndication. The premise for the sitcom involved a newly divorced mother of two daughters who moved to Indianapolis to build a new life. Premiering in the winter of 1976, this show focusing on the trials and tribulations of the single-parent family can, in some sense, be seen as an expression of and response to contemporary social trends, with increased media attention to the rise of households headed by women. (This concern or interest was not limited to television but was also apparent in mass-market magazines, newspaper articles, and so forth.) In the course of the show Ann Romano, the head of the household, had to get a job, raise her children, deal with her former husband's remarriage, and come to grips with her own social life. Thus, the program addressed a wide range of modern social and cultural issues—especially as the children went through high school and grew to adulthood—as the protagonist and her family learned to live "one day at a time."

Throughout this development, however, the traditional nuclear family provided a persistent reference point. From the outset it was figured in symbolic form through the importance of Schneider, the superintendent of the apartment building where Ann Romano lived with her daughters. Although his working-class, "macho" character was often the occasion for humor, he nonetheless assured a male presence in the family. As superintendent, Schneider had a passkey that guaranteed access to the Romano household. This key was a frequent source of narrative contention and humor, but it nevertheless secured freedom of passage into what was otherwise a woman's space. Moreover, as the show developed, Schneider's initially buffoonish flirtations with Ann Romano were at times treated as a more serious and sincere possibility. Over time he indeed became a close friend and father figure to the Romano household. When this relationship was not foregrounded, Ann's boyfriends and the intermittent appearance of her former husband provided a male presence in the household, even if that position was not consistently held by a single individual.

Moreover, Ann's role as a single parent struggling for independent security did not serve as a general narrative model for her daughters. Ann strongly promoted the idea that they should go to college and develop independent skills before attaching themselves to a man. In other words, she urged them not to repeat her fictional past behavior, which comprised the very premise of the show itself. But in the course of the program's history, both daughters married before attending or finishing college. The implication here is that Ann's past behavior—the motive force for the whole fiction, a mistaken course of past action for which each weekly episode was a form of compensation—is somehow eternal. Women's choice of love and marriage as priorities over independent achievement is represented as the most typical behavior, the strongest social model. One woman—here, Ann—might end up divorced and have to struggle to reachieve the middle-class comforts of a former marriage on her own, but this should not become a compelling reason for her daughters to attend college or develop professional skills before marrying. Cutting back across this general pattern of development, the older daughter was portrayed as having an array of marital problems of her own—separating from her husband, reuniting when she discovers that she is pregnant, and finally leaving her husband and child to seek independence. In part, these shifts in her behavior, which included long absences from the show, were explained by external factors of production (well covered in *TV Guide* and other common sources of information about the medium) in terms of the instability of the actress, Mackenzie Phillips, until it was finally revealed that she had problems with drug addiction.

In the final season of the show, Ann Romano herself remarried, this time to her son-in-law's father. Thus, the show progressed to the point of exacerbated familial reformulation, with redoubled family ties. The show ceased production after one season dealing with Ann's readjustments to marriage—with husband and wife working out the details of renewed marital dependence—as if to suggest that it was no longer necessary for her to face life one day at a time. In this way, a program that dealt with an array of social, economic, cultural, and personal issues of the "new" single-parent family was sustained and held in place by constant reference to the traditional nuclear family as the ideological a priori of its overall development.

The ideological meanings and voices produced on television are not unified or monolithic. The result is not that television in episodes or series can mean "anything" you want it to, or has something for everyone. Rather, there is a range of intersecting, and at times even contradictory, meanings through the course of programming offering some things for most people, a regulated latitude of ideological positions meeting the interests and needs of a range of potential viewers. This regulated latitude does not encompass extreme positions and is offered with a strong emphasis on balance and even-handedness to hedge against offending any moderate position. Again, it might be useful to approach this in terms of the ideological problematic that the program constructs for viewers. Within individual programs, between programs and commercials, and across a variety of programs, television is highly fragmented and heterogeneous, allowing for the orchestration of a variety of issues, voices, positions, and messages. None of these, on its own, accounts for the ideology of the medium. Rather, the aim of ideological analysis is to understand their coexistence and contradiction through the medium. In the process of offering a concert of voices, and with its strong links to consumerism, television works to sustain dominant social-cultural ideology, while allowing that this itself involves a series of values and attitudes.

TELEVISION AS A HETEROGENOUS UNITY

The recognition of television's regulated ideological plurality raises the question of viewers—how they engage and are engaged by the medium, and how they are situated in relation to its production of ideological meanings. These issues are more fully addressed in the context of psychoanalytic criticism. With respect to the ideological functioning of the medium, it is nonetheless crucial to understand that, as a site of textual activity, television is the locus of intersection and coexistence of varying narratives, genres, appeals, and modes of address. Viewers consent to watch and to submit to its array of appeals, in exchange for the text and the possibility of identifying particular meanings, mobilizing the voices that seem to speak "to them."

This interpretation, in turn, raises the possibility of subcultural or subversive reading, as particular marginalized or disempowered social groups (women, blacks, gays, and others) find strategies for activating isolated moments within textual flow that pose the possibility of rupturing the dominant ideology. Although this strategy will be further developed in the chapter on British cultural studies, it is useful to bring it up in the context of ideology because it deals with the ways in which individuals might recognize and use the meanings made available through the heterogeneity of television's system of representation.

Feminist approaches to daytime soap opera, for example, have suggested that the traditional villainess transforms feminine weakness into a source of power and strength and offers viewers a figure of female vengeance against patriarchal restraint.[13] To the extent that viewers recognize and identify with her power, the soap opera villainess stands as an emblem or agent of subversion.

Programs such as *Golden Girls* and *Kate and Allie* offer the narrative premise of adult women living together as a family. The female characters in these shows—in couples or in groups—are firmly established as heterosexual, and episodes regularly deal with dating, the desire for male companionship, and past marriages. But at the same time, they validate women's bonding as a form of social stability, a viable and attractive alternative to the traditional family, and even hint at the possibility of lesbian lifestyles—at least as far as possible within dominant ideology. A subcultural reading would emphasize these aspects of the program, a purposeful reading against the grain of narrative events that otherwise conform to normative, dominant heterosexuality. Indeed, such a reading might stress that on a week-to-week basis the narrative privileges women's relations over their inadequate, transient dealings with men. In the same vein, one could look at strong male, familial-type relations in *The A-Team* or *Magnum, P.I.* as offering similar possible appeal as a protohomosexual fantasy. On this basis, one can read these shows in ways that exceed the moderated plurality of dominant social-cultural values and understand their appeal for a particular segment of their audience.

Sub-cultural readings are carried out in the interest of a willful subversion of dominant ideology by social and cultural groups whose interests are not centrally addressed, or are largely ignored, by television's system of representation with its plurality of voices. The claim is not that these shows are intentionally or consciously about alternatives to heterosexuality, or that television in general offers radical representations as an alternative to dominant social-cultural values. Rather, "against the grain" readings are interested in the latent possibility of alternative viewpoints erupting within the multiple strategies of appeal that are normally at work in the medium. For example, in the discussion of *Cagney and Lacey,* we saw how a variety of perspectives and topics were expressed in the show, but ultimately contained by dominant conventions and norms. A subversive reading emphasizes a marginal voice or position and brackets off the dominant context that presumably holds it in place, resisting the pressure of the strategies Fiske and Hartley identify as serving to claw subjects back to the normative viewpoint. These alternative readings become a way of turning the medium on its head, so to speak. They are not limited to issues of sexuality but also include questions of class and ethnicity, and they allow various subcultural audiences to initiate a deconstruction of television's ideology through the medium's own texts.

In part, alternative readings are possible because of the overriding contradiction that characterizes contemporary social practice in general and television in particular. In striving to represent itself as a totality that speaks for and to us all, the medium inevitably raises issues and points to values and ideas that are problematic or disruptive, and that cannot be neatly or easily subsumed in general social consensus. The combined texts of television nevertheless work to hold themselves together as the diversified expression of dominant ideology. This struggle for unity occurs not only at the level of ideas and issues but also at the level of genre and mode of address, as television attempts to fashion a unified "world" out of discontinuous textual fragments. Regularity and repetition are important strategies for ordering the unending flow of television's images and sounds. The same shows with the same characters air at the same time each week; reruns and syndicated repeats provide frequent returns to already-known material; news and talk show hosts are promoted as familiar (even familial) individual personalities. All of these contribute to an overall sense of regularity and stability as part of television's appeal across program flow.

Within this context, the manufacture of the celebrity-personality-as-commodity can be seen as a crucial strategy working to hold the diversity of television's textual flow in place for the viewer—another mechanism for recycling a specific "product" through a system of textual segments. One might point to the figure of Ed McMahon as prototypical in this respect. His prominent status as a nationally recognized figure is anchored in his multiple appearances within television—as Johnny Carson's "second banana" on *The Tonight Show,* as commercial spokesman for a variety of advertised products, as the promoter for the Publisher's Clearinghouse Sweepstakes, and as host of the syndicated *Star Search.* The celebrity presence of Mr. T can similarly be traced through television, as an initial appearance in a "World's Toughest

Bouncer" contest led to a starring role in the film *Rocky III*. After this, he appeared as a guest on *Late Night with David Letterman*, followed by his role on *The A-Team* and the subsequent development of the *Mr. T* Saturday morning cartoon show. This proliferation of roles led to various appearances in made-for-TV movies, variety specials, interview shows, commercials, and so forth, capped by his recent involvement with Hulk Hogan and wrestling on television. Ed McMahon and Mr. T may seem to be extreme and singular examples, but their careers are in fact quite typical of the way the medium deploys familiar individuals. The habitual regularity of the medium is not limited to programs, genres, and schedules, but includes the individuals who populate it.

This sort of proliferation signals the status of the celebrity as a commodity, a figure of circulation, and simultaneously allows viewers to recognize terms of unity in relation to the celebrity persona across a range of genres, programs, and audiences. Although specific commercials, episodes, and programs maintain their integrity and impact as individual texts, they also constantly refer to one another.[14] The recognition and enjoyment of intramedium connections and references at this level means that any given text or textual segment makes sense on its own but also may evoke a second field of referentiality. On the one hand, a text may be provisionally excised from the flow of television and analyzed on its own terms, according to a variety of methods. On the other hand, the rest of the medium becomes the representational context that grounds the program within television as a self-defined textual field. This process of double referentiality is not limited to the deployment of celebrity figures. It is equally apparent in the self-reflexivity and intertextual references engaged in a wide range of programs, episodes, commercials, and so forth, including *Late Night with David Letterman*, *Remington Steele,* and *Moonlighting,* among others.[15]

A full understanding of television's ideological production must take account of this aspect of the meanings generated through the medium. Indeed, in these terms, we might return to the cough medicine commercial discussed at the start of this chapter and consider an additional possible reading: At an extreme one might assume that the ad is hermetically self-referential and that the actor, who is not a doctor, plays a doctor *only* in this ad, and nowhere else on television. In this case the interplay of extratextual and intertextual reference is caught up in a sort of mirror logic, signalled with the declaration, "I'm not a doctor, but . . ." The verbal message implies that the actor plays a doctor in a different television text. But if we assume (even playfully) that he plays a doctor *only in this commercial,* then we may conclude that the ad is obviously referring to itself, and to all of the times it is broadcast. Thus intertextuality becomes self-conscious self-referentiality, as the implied reference to *another* text is actually only a reference to *this* text. Here television exhibits its own fictionality, but in terms that insist that this fictionality exercises affective and intellectual appeal. In this extreme interpretation, television is at once completely artificial and completely meaningful to its viewers. We know that the ad doesn't really convey the voice of medical authority—it is only an actor playing a doctor in a commercial. And the ad tells us that it knows we recognize this artificiality. But we may still follow its lead and buy the product to alleviate a cough, which is also something the ad wants us to do.

Because the commercial has so clearly set up the terms of its functioning within the conventions of the medium, it is easy—even effortless—to watch it and to follow its logic, as long as we already understand the medium's norms, its "natural" practices and strategies. Ideological criticism is precisely concerned with understanding these norms, with the terms in which television presents itself as natural, effortless, and therefore as pleasurable and meaningful to its viewers. Even at its points of minimal referentiality, with its self-reflexive acknowledgement of its own fictionality, this particular commercial—along with many other television texts—fits into the world constructed on television and works to position us as potential consumers in a real marketplace beyond the confines of the television screen. Ideological analysis allows us to understand the strategies and mechanisms of television that produce these paradoxical and contradictory positions of knowledge within contemporary culture.

Finally, ideological criticism is concerned with texts as social processes and as social products. Given television's prominent position in contemporary social life, its dense network of texts, and its pervasive implication in a larger consumer culture, it constitutes a prominent sphere of contemporary ideological practice. It is thus clearly important to subject the medium to ideological investigation. At the same time, these characteristics make the project of ideological analysis a complex task. Because of its fragmentation and heterogeneity, television constantly draws viewers into its "world" of representation, but it may do this in uneven or variable ways. Yet this heterogeneity is not the occasion for limitless perspectives but functions as a limited and regulated pluralism, striving to hold things in balance and to develop its subjects and points of view in relation to normative frames of reference.

Dominant ideological interests may constitute this normative frame and prevail in the last instance. But along the way we are confronted with a variety of issues, ideas, and values that cannot be easily subsumed under the heading of "ruling ideology," which is itself constructed in contradiction. This is further complicated by the fact that, in the current social formation, television itself contributes to, and exists in, highly fragmented and dispersed systems of representation, so that it is difficult to identify a single normative or dominant voice. In the face of this heterogeneity it is all the more crucial to directly confront and analyze the mobilization of multiple perspectives and contradictions through and across the texts that comprise television, to develop our understanding of ideological practice in all its complexity.

NOTES

1. The discussion of Marxist theory developed here is intended as a general and introductory overview. In the process of summary, I inevitably and unfortunately simplify an important and complex body of literature, conflate a broad range of diverse thought, and elide refinements and subtleties within Marxist theory. Rather than attempting to sort out and individually identify the full range of Marxist perspectives in the course of the chapter, I have included key texts on Marxist theories of ideology in the supplemental bibliography.

2. These areas of social practice—economic, political, and ideological—do not exhaust human experience, but designate key arenas within which individuals find their social identity within the social formation. See Louis Althusser, "Ideology and Ideological State Apparatuses," in *Lenin and Philosophy*, trans. Ben Brewster (New York: Monthly Review Press, 1971), pp. 127–86. For further elaboration and discussion, see Rosalind Coward and John Ellis, *Language and Materialism* (London: Routledge and Kegan Paul, 1977), especially pp. 61–92.

3. The specific formulation is paraphrased and borrowed from several theorists who have discussed Althusser's theory of ideology, in particular Stuart Hall, "Signification, Representation, Ideology: Althusser and the Post-Structuralist Debates," *Critical Studies in Mass Communication* 2, no. 3 (June 1985): 103; and Coward and Ellis, *Language and Materialism*, p. 67. Althusser explains

ideology in these terms in "Marxism and Humanism," in *For Marx*, trans. Ben Brewster (New York: Vintage Books, 1970), pp. 221–47.

4. Stuart Hall's "Signification, Representation, Ideology" offers what I would consider a revision or reappraisal of Althusser that draws heavily, but not uncritically, on his theory. A more thoroughgoing critique is to be found in Simon Clarke et al., *One-Dimensional Marxism* (London: Allison and Busby, 1980).

5. Hall, "Signification, Representation, Ideology," p. 105.

6. Ibid.

7. For example, Jane Feuer discusses the importance of a "quality audience" in the development of MTM Enterprises in "MTM Enterprises: An Overview," in *MTM: Quality Television*, ed. Jane Feuer et al. (London: British Film Institute, 1984), pp. 1–31.

8. A variety of alternative perspectives have been developed about how the medium engages its viewers. Some of these are summarized by William Boddy, "Loving a Nineteen-Inch Motorola: American Writing on Television," in *Regarding Television—Critical Approaches: An Anthology*, ed. E. Ann Kaplan, American Film Institute Monograph Series, vol. 2 (Frederick, Md.: University Publications of America, 1983), pp. 1–11. In particular, he provides an overview of the so-called "pessimistic" culture theorists who perceived mass media as organizing popular taste "along the demands of the consumer market" (p. 4) and offering the alienated and fragmented masses of industrial society a false sense of community.

Others have discussed the medium's appeal in terms of its utopian kernel, as it responds to *real* social needs, if in delimited or defined ways. See Hans Magnus Enzensberger, *The Consciousness Industry* (New York: Seabury Press, 1974); and Richard Dyer, *Light Entertainment* (London: British Film Institute, 1973), especially, pp. 39–42.

9. Interestingly enough, with regard to media influences in Webster's dream, the behavior of George and Katherine is clearly modeled on "the millionaire and his wife" from *Gilligan's Island*. George wears yachting clothes and affects the speech of Mr. Howell while Katherine dresses over-formally—an elaborate gown and a fur coat—and behaves like the empty-headed Mrs. Howell.

10. David Morley, *The "Nationwide" Audience: Structure and Decoding* (London: British Film Institute, 1980), p. 139.

11. Ibid.

12. John Fiske and John Hartley, *Reading Television* (London: Methuen, 1978), p. 87. These issues are raised in discussion of the made-for-TV movie in Laurie Jane Schulze, "*Getting Physical*: Text/Context/Reading and the Made-for-Television Movie," *Cinema Journal* 25, no. 2 (Winter 1986): 35–50.

13. Tania Modleski, *Loving with a Vengeance: Mass-Produced Fantasies for Women* (London: Methuen, 1982), pp. 95–98.

14. For a more systematic analysis of how television constructs these unities and continuities as a mechanism of viewer engagement, see Mimi White, "Crossing Wavelengths: The Diegetic and Referential Imaginary of American Commercial Television," *Cinema Journal* 25, no. 2 (Winter 1986): 51–64.

15. See White, "Crossing Wavelengths." Also, Jane Feuer discusses self-

reflexivity in MTM programs in "The MTM Style," in *MTM: Quality Television*, pp. 32–60.

FOR FURTHER READING

The suggestions for further reading are divided into four broad areas, beginning with Marxist theory of ideology and concluding with examples of ideological analyses of television. However, in the process of organizing particular selections, I have not always maintained firm boundaries. For example, I have included an article on television by Theodor W. Adorno in section 1, along with other readings by the Frankfurt School theorists. Similarly, Raymond Williams's book on television appears in section 2, along with other works by Williams on culture and society.

MARXIST THEORY OF IDEOLOGY

One of the earliest elaborations of ideology in the writings of Marx is Karl Marx and Frederick Engels, *The German Ideology* (Moscow: Progress Publishers, 1976), part 1. Also see the collection of Marx and Engels, *On Literature and Art* (New York: International General, 1973).

The work of Antonio Gramsci is available in *Selections from the Prison Notebooks*, ed. and trans. Quentin Hoare and Geoffrey Nowell-Smith (New York: International Publishers, 1971) and *Selections from Cultural Writings*, trans. William Boelhower (Cambridge, Mass.: Harvard University Press, 1985).

Louis Althusser develops his theory of ideology in a number of essays in *For Marx*, trans. Ben Brewster (New York: Vintage Books, 1970) and in *Lenin and Philosophy*, trans. Ben Brewster (New York: Monthly Review Press, 1971).

The Althusserian position on ideology is discussed and elaborated in relation to semiotics, psychoanalysis, and the theory of the subject in Rosalind Coward and John Ellis, *Language and Materialism* (London: Routledge and Kegan Paul, 1977). A critique of the Althusserian position, in particular in relation to understanding culture, is offered by Simon Clarke et al., *One-Dimensional Marxism* (New York: Allison and Busby, 1980).

The Frankfurt School offers Marxist perspectives on sociology and culture that were not developed in this chapter. Their contributions to Marxist theories of mass culture have been significant. In particular, see Max Horkheimer and Theodor W. Adorno, *Dialectic of Enlightenment*, trans. John Cumming (New York: Seabury Press, 1972), especially "The Culture Industry: Enlightenment or Mass Deception." Also see Andrew Arato and Eike Gebhardt, eds., *The Essential Frankfurt School Reader* (New York: Urizen Books, 1978); and Theodor W. Adorno, "Television and the Patterns of Mass Culture," in *Television: The Critical View*, ed. Horace Newcomb, 1st ed. (New York: Oxford University Press, 1976), pp. 239–59. (Note: The Adorno essay is not included in more recent editions of the Newcomb anthology.)

IDEOLOGY IN LITERATURE AND THE ARTS

Much of the discussion and debate over theories of ideology and culture have been developed in relation to literary studies and the arts. Some of this work may be of theoretical interest to those who wish to pursue issues in ideological criticism.

Raymond Williams is a crucial figure, elaborating sociological perspectives on literature and culture within the context of Marxist theory. In *Keywords* (New York: Oxford University Press, 1976) he traces key terms and concepts in culture and society. Also see *Culture* (Glasgow: Fontana, 1981) and *Problems in Materialism and Culture* (London: Verso, 1980). Williams also wrote one of the earliest books on television in the tradition of British cultural studies, *Television: Technology and Cultural Form* (New York: Schocken Books, 1975).

Within the context of literary theory, Althusserian perspectives are developed in Pierre Macherey, *A Theory of Literary Production*, trans. Geoffrey Wall (London: Routledge and Kegan Paul, 1978); and in Fredric Jameson, *The Political Unconscious: Narrative As a Socially Symbolic Act* (Ithaca, N.Y.: Cornell University Press, 1981).

Approaches to art as a social product, including Marxist theories of ideology, can be found in Janet Wolff, *The Social Production of Art* (New York: New York University Press, 1984).

IDEOLOGY AND MASS CULTURE

Useful essays on ideology and culture, including film, television, and mass media, are collected in Michele Barrett et al., eds., *Ideology and Cultural Production* (New York: St. Martin's Press, 1979). A collection of essays from the Birmingham Centre for Contemporary Cultural Studies, including Marxist approaches to culture, is Stuart Hall et al., eds., *Culture, Media, Language* (London: Hutchinson, 1980). Media systems, institutional analyses, and the mediation of culture in particular texts are all covered in James Curran et al., eds., *Mass Communication and Society* (Beverly Hills, Calif.: Sage, 1979). Gaye Tuchman, ed., *The TV Establishment: Programming for Power and Profit* (Englewood Cliffs, N.J.: Prentice-Hall, 1979) offers essays on media structures and practices based on the reflection hypothesis that the content and structure of the media reflect social values and needs.

A theoretical discussion of the arts and mass culture within the Marxist tradition, including extensive discussion and critique of Althusser, is available in Terry Lovell, *Pictures of Reality* (London: British Film Institute, 1980).

A summary and overview of different methodological approaches to culture in the Marxist tradition is provided by Lawrence Grossberg, "Strategies of Marxist Cultural Interpretation," *Critical Studies in Mass Communication* 1, no. 4 (December 1984): 392–421.

Two articles on Marxism in relation to cultural studies and the media are in *Film Reader* no. 5 (1982): Nicholas Garnham, "Film and Media Studies: Reconstructing the Subject," pp. 177–83; and Terry Lovell, "Marxism and Cultural Studies," pp. 184–91.

In *Covering Islam* (New York: Pantheon Books, 1981) Edward W. Said discusses the Western, and specifically American, construction of "Islam" as an ideological concept, with particular attention to the media's contribution to this process.

IDEOLOGICAL ANALYSIS OF TELEVISION

The following books include ideological perspectives in the context of more general consideration of particular programs, genres, or issues related to television programming: Robert C. Allen, *Speaking of Soap Operas* (Chapel Hill: University of North Carolina Press, 1985); Ien Ang, *Watching "Dallas": Soap Opera and the Melodramatic Imagination*, trans. Della Couling (London: Methuen, 1985); John Hartley, *Understanding News* (London: Methuen, 1983); and Philip Schlesinger et al., *Televising "Terrorism"* (London: Comedia Series, no. 16, 1983).

A number of the essays in E. Ann Kaplan, ed., *Regarding Television—Critical Approaches: An Anthology*, American Film Institute Monograph Series, vol. 2 (Frederick, Md.: University Publications of America, 1983) emphasize the ideological work of particular programs or genres.

The British Film Institute Television Monograph series offers a number of studies that include considerations of ideological issues. Although they emphasize British television, the theoretical perspectives and critical methods they engage might be useful. For example, Richard Dyer, *Light Entertainment* (London: British Film Institute, 1973) discusses variety/music shows; Colin McArthur, *Television and History* (London: British Film Institute, 1978) addresses the construction of history in television programs. Essays on television from *Screen* and the BFI monograph series are collected in Tony Bennett et al., eds., *Popular Television and Film: A Reader* (London: British Film Institute/Open University Press, 1981), pt. 2: The Discourses of Television; and pt. 4: History, Politics, and Classical Narrative.

There is a growing body of articles included in journals and anthologies that offer ideological analyses of television programs. The following list is hardly exhaustive but tries to cover articles on a variety of programs and genres.

On game shows: Adam Mills and Phil Rice, "Quizzing the Popular," *Screen Education* 41, no. 41 (Winter/Spring 1982): 15–25; John Tulloch, "Gradgrind's Heirs—the Quiz and the Presentation of 'Knowledge' by British Television," *Screen Education*, no. 19 (Summer 1976): 3–13.

On news: Cary Bazalgette and Richard Paterson, "Real Entertainment: The Iranian Embassy Siege," *Screen Education* 37 (Winter 1980/81): 55–67; Ian Connell, "Television, News, and the Social Contract," *Screen* 20, no. 1 (Spring 1979): 87–107.

On dramatic series and programs: Susan Boyd-Bowman, "*The Day After*: Representations of the Nuclear Holocaust," *Screen* 25, no. 4/5 (July/October

1984): 71–97; Jane Feuer, "Melodrama, Serial Form, and Television Today," *Screen* 25, no. 1 (January/February 1984): 4–16; John Fiske, "Popularity and Ideology: A Structuralist Reading of *Dr. Who*," in *Interpreting Television: Current Research Perspectives*, ed. Willard D. Rowland, Jr., and Bruce Watkins (Beverly Hills, Calif.: Sage, 1984): pp. 165–98; Todd Gitlin, "Prime Time Ideology: The Hegemonic Process in Television Entertainment," *Social Problems* 26, no. 3 (1979): 251–66; Laurie Jane Schulze, "*Getting Physical*: Text/Context/Reading and the Made-for-Television Movie," *Cinema Journal* 25, no. 2 (Winter 1986): 35–50; Cathy Schwichtenberg, "*The Love Boat*: The Packaging and Selling of Love, Heterosexual Romance, and the Family," *Media, Culture, and Society* 6, no. 3 (July 1984): 301–11; Ellen Seiter, "The Hegemony of Leisure: Aaron Spelling Presents *Hotel*," in *Television in Transition*, ed. Phillip Drummond and Richard Paterson (London: British Film Institute, 1985), pp. 135–45.

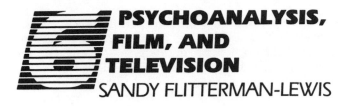

PSYCHOANALYSIS, FILM, AND TELEVISION
SANDY FLITTERMAN-LEWIS

After a day's work at the film studio, Alfred Hitchcock used to doze off in front of the TV screen; "Television," he said, "was made for that purpose." For film theorists, psychoanalysis has provided a useful way of discussing our relationship with the cinema. It has done this primarily through an analogy between film and that product of slumber, the dream—tracing the relationship between films themselves and the dream-work, that unconscious process of transformation that permits us to relate "stories told in images" to ourselves while we sleep. But if the dreamer and the film spectator are kindred spirits in some ways, what kinds of conclusions can we draw when we apply this analogy to the study of television, a medium whose very techniques and processes, while similar in some ways to film, are vastly different in crucial ways? In what follows I will discuss the principles of psychoanalytic criticism as they have developed in film studies, the main features that differentiate film from television, and, finally, the usefulness of applying the tools of psychoanalytic criticism developed in film studies to television, through a discussion of the representative example of soap opera—considered by many to be the "quintessential televisual form." However, from the very outset it is important to emphasize that cinema and television are two completely distinct media; as textual systems, and in the manner by which we engage with them as viewers, film and television are profoundly different. The conditions that produce visual/auditory images, and that shape our viewing experience in the cinema, are simply not the same when we watch TV. For this reason, where psychoanalysis is concerned, there can be no simple exchange of method from one medium to the other. Rather, what the psychoanalytic approach might provide, in its application to television studies, is the definition and description of an entirely new type of social subject, part viewer, part consumer—the "tele-spectator" (to use French filmmaker Jean-Luc Godard's evocative term).

PSYCHOANALYSIS AS A CULTURAL THEORY

In order to analyze the ways in which this *different* TV spectator is constructed and engaged, I will begin by summarizing the basic tenets of psychoanalysis. (This will necessarily require a certain amount of oversimplification on my part, for the argument is complex and fairly resistant to summary. What I intend here is simply to trace the broad outlines of psychoanalytic theory so that its relation to a critical under-standing of both film and television will become clear; readers who would like to pursue this line of argument in depth should consult the bibliography for further reading.) Psychoanalysis, as a theory of hu-man psychology, describes the ways in which the small human being comes to develop a specific personality and sexual identity within the larger network of social relations called culture. It takes as its object the mechanisms of the unconscious—resistance, repression, sexu-ality, and the Oedipal complex—and seeks to analyze the fundamental structures of desire that underlie all human activity. For Freud, who discovered and theorized the unconscious, human life is dominated by the need to repress our tendencies toward gratification (the "pleasure principle") in the name of conscious activity (the "reality principle").[1] We come to be who we are as adults by way of a massive and intricate repression of those very early, very intense expressions of libidinal (sexual) energy. The "unconscious" is what Freud designates as that place to which unfulfilled desires are relegated; as such, it has been referred to as that "other scene" where the "drama of the psyche" is played out. In other words, beneath our conscious, daily social interac-tions there exists a dynamic, active play of forces of desire that is in-accessible to our rational and logical selves.

The unconscious, however, is not simply a ready-and-waiting place for repressed desire—it is *produced* by the very act of repression. In describing the process by which the unconscious is formed, Freud takes the hypothetical life of the infant as it develops from an entity entirely under the sway of libidinal gratifications to an individual capable of establishing a position in a social world of men and women. In this way, Freud's theory of the human mind becomes not simply a parable of individual development, but a general model for the way all of human culture is structured and organized. One of Freud's major contributions to the theory of human personality was his discovery of "infantile sexuality"—there is eroticism in the earliest of our childhood experiences. From the very first moment in an infant's life, the small organism strives for satisfaction of those biological needs (food, warmth, and so on) that can be designated as instincts for self-preservation. Yet at the same time, this biological activity also produces experiences of intense pleasure (sensuous sucking at the breast, a complex of satisfying feelings associated with warmth and holding, and the like). For Freud, this distinction indicates the emergence of sexuality; desire is born in the first separation of the biological instinct from the sexual drive. Importantly, the element of fantasy is already present, for all future yearnings for milk by the infant will be marked by a need to recover that *totality* of sensations that goes beyond the mere satisfaction of hunger. In other words, there is a process of hallucinating—a *fantasmatic* process—going on; each time the child cries for milk, we can say that the child is actually crying for "milk" (milk-in-quotes)—that hallucinated image of the bonus of satisfaction that came when the need of hunger was fulfilled.

As the child grows, there is a gradual organization of the libidinal drives that, while still centered on the child's own body, channels sexuality toward various objects and aims. The first phase of sexual life is associated with the drive to incorporate objects (the oral stage); in the second, the anus becomes the erotogenic zone (the anal stage); and in the third, the child's libido is focused on the genitals (the phallic stage). What is important here is that the child, not yet having a centered self (an ego, an identity) nor being able to distinguish between itself and the outer world, is like a field across which the libidinal energy of the drives plays.

At this point Freud introduces the Oedipal complex, a decisive moment in the child's development, for it defines the individual's emergence into sexed selfhood. In the pre-Oedipal stages, both male and female child are in a dyadic relation with the mother; with the Oedipal moment, this two-term relation becomes three, and a triangle is formed by the child and both parents. The parent of the same sex becomes a rival in the child's desire for the parent of the opposite sex. The boy gives up his incestuous desire for the mother because of the threat of punishment by castration perceived to come from the father; in so doing, he *identifies* with his father (symbolically becomes him) and prepares to take his position of a masculine role in society. The forbidden desire for the mother is driven into the unconscious, and the boy will accept substitutes for the mother/desired object in his future as an adult male. For the female, the Oedipal moment is not one of threat, but of realization—she recognizes that she has *already* been castrated, and, disillusioned in the desire for the father, reluctantly identifies with the mother. In addition, the Oedipal complex is far more complicated for the girl, who must change her love object from mother (the first object for both sexes) to father, whereas the boy can simply continue loving the mother.

Such schematizing makes the claims against Freud's sexism even more evident, and one could write a whole volume on the subject. For the moment, I simply wish to describe the general outlines of the theory, pointing out that Freud did not create, but was simply describing, the mechanisms prevalent in the patriarchal society in which we live. What is relevant for this essay, however, is the work of the unconscious, the production of fantasy, and the erotic component of desire in all of our activities (including watching film and TV). In discussing the Oedipal moment, we should remember that these are *symbolic* structures which take place at the level of the unconscious rather than of felt experience. While we might remember feelings of hostility or intense love for one parent, we cannot remember the Oedipal situation as such, for it is precisely because of *repression* that these experiences become part of our unconscious psychic make-up. The important point here is that the Oedipus complex signals the transition from the pleasure principle to the reality principle, from the familial order to society at large. The threat of castration and the Oedipus complex are the symbolic imposition of a culture's rules—they represent the law, morality, conscience, authority, etc. Freud uses this schema to describe the processes by which the child develops a unified sense of self (an ego) and takes up a particular place in the cultural networks of social, sexual, and familial relations.

For Freud, the individual (or subject) who emerges from this process is irrevocably split between two levels of being—the conscious life of the ego, or self, and the repressed desires of the unconscious. This unconscious is formed by repression, for it is the guilty desires, forced down below the surface of conscious awareness, that cause it to come into being. Thus it is radically distinct from rational conscious life—it is utterly *other*, strange, illogical, and contradictory in its instinctual play of the drives and ceaseless yearning for gratification. Freud says that dreams are the "royal road to the unconscious." This is because dreams are actually symbolic fulfillments of unconscious wishes. (The

Disney song, "A Dream Is a Wish Your Heart Makes," was not too far off.) In order for the unconscious subject to produce a dream—a symbolic "text" that can be understood through a process of decipherment, unravelling the various threads of dream-imagery to get to the core, the "dream-wish" itself—the unconscious engages in something called the *dream-work*. Various operations such as *condensation* (in which a whole range of associations can be represented by a single image), *displacement* (in which psychic energy is transferred from something significant to something banal, conferring great importance on a trivial item), *conditions of representability* (in which it becomes possible for certain thoughts to be represented by visual images), and *secondary revision* (in which a logical, narrative coherence is imposed on the stream of images) combine to transform the raw materials of the dream (bodily stimuli, things that happened during the day, dream-thoughts) into that hallucinatory "visual story" which is the dream itself.

With the transforming work of the dream as an example, we can see that the workings of the unconscious find no *direct* expression in conscious life (because these workings are the result of an initial repression). However, the complicated pathways between conscious activity and unconscious desire are made evident through the vehicle of language. As dreams, neuroses (the result of an internal conflict between a defensive ego and unconscious desire), slips of the tongue, failures of memory, and jokes and puns indicate, unconscious wishes and desires—with a logic of their own—underlie even the most apparently "innocent" activity. Even the simple acts of filmgoing or watching TV are shaped by unconscious desires. This fact implies that there can never be a one-to-one relationship between language and the world; meaning always *exceeds* its surface, and things do not always "mean" what they appear to. We can never say with any certainty that the speaking subject says exactly what it means or means what it says; we can never possess the "full" meaning of any of our actions.

Thus we know of the existence of the unconscious when it "speaks" to us through the language of dreams, neuroses, and the like. This emphasis on expression has led French psychoanalyst Jacques Lacan to say that the unconscious is "structured like a language." Lacan is credited with reinterpreting Freud in the context of structural linguistics, and it is the work of Lacan that psychoanalytic film theory is based on. Because of his emphasis on language, Lacan rereads the Oedipal complex along these lines: the child moves out of the pre-Oedipal unity with the mother not only through fear of castration, but through the acquisition of language as well. Thus the moment of linguistic capability (the ability to speak, and to distinguish a speaking self) is the moment of one's insertion into a social realm (a world of adults and verbal exchange). All of us learn to speak in the language and customs of our particular culture; Lacan inverts this to say that we are in fact *spoken* by the culture itself. Our sense of self is formed through the perception and language of others, and it is even at the deepest levels of the unconscious that this takes place.

Lacan develops a theory in which the questions of the human subject (individual), its place in society, and its relationship to language are all interconnected. He charts the development of the self and the formation of the psyche in terms of psychoanalytic "registers" that are roughly equivalent to Freud's pre-Oedipal and Oedipal phases. In what Lacan calls the "Imaginary" realm, the child's first development of an ego—an integrated self-image—begins to take place. It is here in the "Mirror Phase," Lacan says, that this ego comes into being through the infant's identification with an image of its own body. Between the ages of six and eighteen months, the human infant is physically uncoordinated; it perceives itself as a mass of disconnected, fragmentary movements. It has no sense that the fist which moves is connected to the arm and body, and so forth. When the child sees its image (for example, in a mirror—but this can also be the mother's face, or anyone perceived as whole), it mistakes this unified, coherent shape for a superior self. The child *identifies* with this image (as both reflecting the self, and as something *other*), and finds in it a kind of satisfying unity that it cannot experience in its own body. The infant internalizes this image as an "ideal ego," and this process forms the basis for all later identifications, which are imaginary in principle. Simply put, in order for communication to occur at all, we must at some level be able to say to each other, "I know how you feel." The ability to temporarily—and imaginatively—*become* someone else is begun by this original moment in the formation of the self.

Lacan's "Symbolic" register is roughly equivalent to the Oedipal process and connotes the realm of all discourse and cultural exchange. A third term, symbolized by the father and signifying the Law (of culture) disrupts the harmony of the dual relation in the Imaginary. The Symbolic Order is the realm of preestablished social structures (Lacan uses language as his model), such as the taboo on incest, that regulates relations of marriage and exchange. In this schema, the figure of the father represents the fact that a wider familial and social network exists, and that the child must seek a position in that context. The child must go beyond the imaginary identifications of the dual realm in which the distinction between "me/you" is always blurred, to take a position as someone who can designate himself as an "I" in a world of adult thirds ("he," "she," and "it"). The appearance of the father thus prohibits the child's total unity with the mother, and, as noted before, causes desire to be repressed in the unconscious. Lacan's contribution to psychoanalytic theory involves his rethinking of the Oedipal process in terms of language—when we enter the Symbolic Order we enter language/culture itself.

But because, as we have seen, the unconscious is the radically different site of repression, we are never entirely in control of our meanings. While in conscious life, we have some idea of ourselves as reasonably unified and coherent, this self-perception is in some sense *fictive*. The ego is simply a function or "effect" of that which is always beyond our grasp in the unconscious. When we speak there is never simply a complete, obvious, or logical meaning. This is what is meant when Lacan says that the subject is always split in language. You the subject, as in the subject of a sentence, always take up a somewhat arbitrary position when speaking. The pronoun "I" *stands in* for the ever-elusive subject, the speaking self. When I say "I am lying to you," the "I" in the sentence is fairly stable and coherent; but the "I" that pronounces the sentence (and throws its truthfulness into question to boot) is an always-changing, shifting force. For the sake of understanding, the "I" of the sentence and the one who produces/pronounces it are put into a unity that is of an imaginary kind. Thus, there is a certain level of illusion about identity; we stabilize the shifting that happens in speaking in order to make communication possible.

Lacan's work demonstrates an alliance between language, the unconscious, parents, the symbolic order, and cultural relations. Language is what internally divides us (between conscious and unconscious), but it is also that which externally joins us (to others in culture). By reinterpreting Freud in linguistic terms, Lacan emphasizes the relations between the unconscious and human society. We are all bound to culture by relations of desire; language is both that which speaks from deep within us (in patterns and systems that preexist our birth), and that which we speak in our continual network of relations with others. It is in this sense that psychoanalysis can be interpreted as a social theory.

PSYCHOANALYSIS AND FILM STUDIES

Early in "The Imaginary Signifier," his classic study of film spectatorship, Christian Metz poses a founding question: "What contribution can . . . psychoanalysis make to the study of the cinematic signifier?"[2] In other words, how can the theory of the unconscious help us to understand what happens when we watch a film—how we interact with it, how it creates its meanings, what we come away with? This question echoes throughout Metz's work, emphasizing that: 1) we can't discuss the film spectator without taking the processes of the unconscious into account; and 2) psychoanalysis brings something to the study of film that other types of study leave out. This is because a psychoanalytic approach to the cinema shifts its emphasis away from the film itself—that discrete, formal entity on the screen—toward the spectator, or more precisely, toward the spectator-text relations that are central to the processes of meaning-production in film.

Film theory looks to psychoanalysis to understand why the cinema so immediately became such a pervasive and powerful social institution. For this reason, it is at the level of the cinema's institutional form that Metz first stakes his argument for psychoanalysis; he speaks of the "dual kinship" between the psychic life of the spectator and the financial or industrial mechanisms of the cinema. The cinema reactivates—in ways that are pleasurable—those very deep and globally structuring processes of the human psyche.

> The cinematic institution is not just the cinema industry (which works to fill cinemas, not to empty them). It is also the mental machinery—another industry—which spectators "accustomed to the cinema" have internalised historically, and which has adapted them to the consumption of films. (The institution is outside us and inside us, indistinctly collective and intimate, sociological and psychoanalytic, just as the general prohibition of incest has as its individual corollary the Oedipus complex . . . or perhaps . . . different psychical configurations which . . . *imprint* the institution

in us in their own way.) The second machine, ie, the social regulation of the spectator's metapsychology, like the first, has as its function to set up good object relations with films. . . . The cinema is attended out of desire, not reluctance, in the hope that the film will please, not that it will displease. . . . [T]he institution as a whole has filmic pleasure alone as its aim.[3]

Unlike the models of mass audience offered by empirical or sociological approaches to the cinema ("real" people who go to movies), and unlike the notion of a consciously aware viewer provided by formalist approaches (people have conscious artistic ideas about what they see), psychoanalytic film theory discusses film spectatorship in terms of the circulation of desire. That is, it considers both the viewing-state and the film-text alike as, in some way, mobilizing the structures of unconscious fantasy. More than any other form, the cinema is capable of actually reproducing or approximating the structure and logic of dreams and the unconscious. From Freud, we know that "fantasy" refers to the fulfillment of a wish by means of the production of an *imaginary scene* in which the subject/dreamer, whether depicted as present or not, is the protagonist. In the words of French post-Freudians Jean Laplanche and Jean-Baptiste Pontalis, "[U]nconscious ideas are organized into phantasies or imaginary scenarios to which the instinct becomes fixated and which may be conceived of as true *mises-en-scène* [stagings/performances] of desire."[4] The important point here is that psychoanalytic film theory emphasizes the notion of *production* in its description, focusing on the ways in which the viewer is positioned, by means of a series of "lures," as the *desiring producer* of the cinematic fiction. According to this idea, then, when we watch a film we are somehow *dreaming* it as well; our unconscious desires work in tandem with those that generated the film-dream.

This implies that the spectator is actually a central part of the entire cinema-machine (an apparatus that goes beyond mere films themselves to the whole range of operations involved in their production and consumption). Crucial to this idea is the concept of the *cinematic apparatus*, which the important work of Jean-Louis Baudry ("The Ideological Effects of the Basic Cinematographic Apparatus" and the more fully psychoanalytic "The Apparatus: Metapsychological Approaches to the Impression of Reality in the Cinema") has defined.[5] The cinematic apparatus is thus a complex, interlocking structure involving: 1) the technical base (specific effects produced by the various components of the film equipment, including camera, lights, film, and projector); 2) the conditions of film projection (the darkened theater, the immobility implied by the seating, the illuminated screen in front, and the light beam projected from behind the spectator's head); 3) the film itself, as a "text" (involving various devices to represent visual continuity, the illusion of real space, and the creation of a believable impression of reality); and 4) that "mental machinery" of the spectator (including conscious perceptual as well as unconscious and preconscious processes) that constitutes the viewer as a desiring subject. From this it should be clear that there are both technological *and* libidinal/erotic components which intersect to form the cinematic apparatus as a whole. And at the very center of the cinematic apparatus, there is the spectator, for without this viewing subject, the entire mechanism would cease to function.

But can we say with any certainty *who* this spectator is? What, exactly, is this "fictive participation" on the part of the spectator, and what specific psychoanalytic processes are engaged? The first thing we can note about the cinema spectator is his/her capacity for belief—the cinema spectator is first and foremost a *credulous* one. Drawing his discussion in part from the work of psychoanalyst Octave Mannoni, Metz tells us that belief in the cinema involves a basic process of denial or disavowal.[6] Behind every incredulous spectator (who *knows* the events taking place on the screen are fictional) lies a credulous one (who nevertheless *believes* these events to be true); the spectator thus *disavows* what s/he knows in order to maintain the cinematic illusion. The whole effect of the film-viewing situation turns on this continual back-and-forth of knowledge and belief, this split in the consciousness of the spectator between "I know full well . . ." and "But, nevertheless . . . ," this "no" to reality and "yes" to the dream. The spectator is, in a sense, a double-spectator, whose division of the self is uncannily like that, as we have seen, between conscious and unconscious. So even at the very basic level of belief in the cinematic fiction, something akin to unconscious desire is at work.

Now we come to what is perhaps the trickiest notion in psychoanalytic film theory's conception of film spectatorship. For film theory sees the viewer not as a person, a flesh-and-blood individual, but as an *artificial construct*, produced and activated by the cinematic apparatus. The spectator is discussed as a "space" that is both "productive" (as in the production of the dream-work) and "empty" (anyone can occupy it); the cinema in some sense *constructs* its spectator through what is called the "fiction-effect." There are certain conditions that make film viewing similar to dreaming: we are in a darkened room, our motor activity is reduced, our visual perception is heightened to compensate for our lack of physical movement. Because of this, the film spectator enters a "regime of belief" (where everything is accepted as real) that is like the condition of the dreamer. The cinema can achieve its greatest power of fascination over the viewer not simply because of its impression of reality, but more precisely because this impression of reality is intensified by the conditions of the dream. The cinema thus creates an impression of reality, but this is a total *effect*—engulfing and in a sense "creating" the spectator—which is much more than a simple replica of the real.

Psychoanalytic film theory goes to great lengths to distinguish between the real person and the film viewer, drawing on operations of the unconscious for its description. Three factors go into the psychoanalytic construction of this viewer: 1) regression, 2) primary identification, and 3) the concealment of those "marks of enunciation" that stamp the film with authorship. First, those conditions of the dream state that we've just discussed also produce what Baudry calls "a state of artificial regression."[7] The totalizing, womblike effects of the film-viewing situation represent, for him, the activation of an unconscious desire to return to an earlier state of psychic development, one before the formation of the ego, in which the divisions between self and other, internal and external, have not yet taken shape. For Baudry, this condition in which the subject cannot distinguish between perception (of an actual thing) and representation (an "image" that stands in for it) is like the earliest forms of satisfaction of the infant in which, as you remember, the boundaries between itself and the world are confused. Baudry says that the cinema situation reproduces the *hallucinatory* power of a dream because it turns a perception into something that looks like a hallucination. But he says that there is an important difference. Where Freud says that the dream is a "normal hallucinatory psychosis" of every individual, Baudry points out that film offers an "artificial psychosis without offering the dreamer the possibility of exercising any kind of immediate control."[8]

Yet in order for the slippage from dreamer to viewer to occur—a slippage which defines the peculiar situation of cinema viewing—in order for the film spectator to actually become the subject of someone else's dream (the film), a situation must be produced in which the viewer is "more immediately vulnerable and more likely to let his own fantasies work themselves into those offered by the fiction machine."[9] This has already been prepared for by the heightened receptivity (a state something like the suggestibility of hypnosis) produced by the "artificial regression" of the fiction-effect.

Metz defines primary cinematic identification as the spectator's identification with the act of looking itself: "I am *all-perceiving* . . . the *constitutive* instance, in other words, of the cinematic signifier (it is I who make the film). . . . [T]he spectator *identifies with himself*, with himself as a pure act of perception (as wakefulness, alertness): as condition of possibility of the perceived and hence as a kind of transcendental subject, anterior to every *there is*."[10] This type of identification is considered *primary* because it is what makes all secondary identifications with characters and events on the screen possible. This process, both perceptual (the viewer sees the object) and unconscious (the viewer participates in a fantasmatic or imaginary way), is at once constructed and directed by the look of the camera and its stand-in, the projector. From a look that proceeds from the back of the head, then, "precisely where fantasy locates the 'focus' of all vision," . . . the spectator is given that illusory capacity to be everywhere at once, that power of vision for which the cinema is famous.[11] Baudry describes this arrangement in a slightly more technological way: "[T]he spectator identifies less with what is represented, the spectacle itself, than with what stages the spectacle, makes it seen, obliging him to see what it sees; this is exactly the function taken over by the camera as a sort of relay."[12]

Metz says that this type of identification is possible because the viewer has already undergone that formative psychic process called the Mirror Phase (discussed earlier). The film viewer's fictional participation in the unfolding of events is made possible by this first experience of the subject, that early moment in the formation of the ego when the small infant begins to distinguish objects as different from itself. What links this process to the cinema is the fact that it occurs in terms of visual images—what the child *sees* at this point (a unified image that is distanced and objectified) forms how s/he will interact with others at later stages in life. The *fictive* aspect is also crucial here—the perception of that "other" as a more perfect self is also a *mis*perception. We should remember that the notion of the self as fully conscious, coherent, and in command of its meanings is a fictional construct—the unconscious tells us this is so.

Film theory has been quick to appreciate the correspondence between the infant in front of the "mirror" and the spectator in front of the screen, both being fascinated by and identifying with an imaged ideal, viewed from a distance. This early process of ego construction, in which the viewing subject finds an identity by absorbing an image in a mirror, is one of the founding concepts in the psychoanalytic theory of cinema spectatorship and the basis for its discussion of primary identification. Part of the cinema's fascination, then, comes from the fact that while it allows for the temporary loss of ego (the film spectator "becomes" someone else), it simultaneously reinforces the ego (through the processes discussed). In a sense, the film viewer both loses him/herself, and refinds him/herself—over and over—by continually reenacting the first fictive moment of identification and establishment of identity.

You will remember that the third element in this construction of the cinematic viewer (after regression and primary identification) has to do with "authorship" and its effacement. In our discussion of the viewer as dreamer, I noted that a number of conditions combine to give the spectator the impression that it is he or she who is dreaming the images and situations (the "ideational representatives" of Freud's dream-work) that appear on the screen. Dream and fantasy have this in common with fiction: they are all imaginary productions that have their source in unconscious desire. Freud is very concise when he summarizes this function of the desiring subject: "His Majesty the Ego, the hero of all day-dreams and all novels."[13] But something must happen to hide the "real" dreamer—the implied author of the film— from view; the viewer must be made to forget that a fiction is being watched, a fiction which has, in a sense, come from another source of desire.

For, while all fantasies originate from the subject who produces them, film obviously involves a more complicated process, one in which the unconscious desire—of both filmmaker and spectator—is but one element in the complex operations of technology and text. I pointed out earlier that the viewer's position is produced as an "empty space" so that the viewer is more susceptible to having his/her own fantasies interact with the film. The primary way this is achieved is by shifting the terms of what film theory calls the "system of enunciation." The concept is borrowed from structural linguistics and (if you remember our earlier example) implies the position of the speaking subject. In other words, in every verbal exchange there is both the statement (what is said, the language itself) and the process that produces the statement (how something is said, from what position). According to French linguist Emile Benveniste, "what characterizes enunciation in general is the emphasis on the discursive relationship with a partner, whether it be real or imagined, individual or collective." [14]

Film theory applies this concept to the cinema. In every film there is always a place of enunciation—a place from which the cinematic discourse proceeds. This is theorized as a *position,* not to be confused with the actual individual, the filmmaker. French film theorist Raymond Bellour defines this system of cinematic enunciation in the following way: "A subject endowed with a kind of infinite power, constituted as the place from which the set of representations are ordered and organized, and toward which they are channeled back. For that reason, this subject is the one who sustains the very possibility of any representation." He expands this, in discussing the films of Hitchcock, by remarking on how one can easily move, through the pathways of different characters' points of view, "to a central point from which all these different visions emanate: the place, at once productive and empty, of the subject-director." [15]

In another essay about film viewing ("History/Discourse: A Note on Two Voyeurisms"), Metz connects the process of enunciation to *voyeurism,* the erotic component of seeing that founds the cinema. [16] In psychoanalytic terms, voyeurism applies to any kind of sexual gratification obtained from vision, and is usually associated with a hidden vantage point. Metz theorizes this libidinal visual energy in order to show how the space of cinematic *enunciation* becomes the position of cinematic *viewing:* "If the traditional film tends to suppress all the marks of the subject of enunciation, *this is in order that the viewer may have the impression of being that subject himself,* but an empty, absent subject, a pure capacity for seeing." [17] In other words, in order for the cinematic fiction to both produce and maintain its fascinating hold on the spectator, it must appear as if the screen images are the expressions of the spectator's own desire. Or rather, as Bertrand Augst concisely describes it, "The subject-producer must disappear so that the subject-spectator can take his place in the production of the filmic discourse." [18]

For his model of the cinema, Metz transforms the linguistic emphasis of enunciation into a concept of the enunciator as "producer of the fiction," indicating that process by which every filmmaker organizes the image flow, choosing and designating the series of images, organizing the diverse views that make up the relay between the one who looks (the camera, the filmmaker) and what is being looked at (the scene of the action). Metz maintains that one of the primary operations of the classical narrative film (what distinguishes it as "classical" in fact), is the effacement or hiding of these "marks of enunciation" which point to this work of selecting and arranging shots, and in a sense, reveal the filmmaker's hand. This is the famous "invisible editing" of Hollywood cinema. The work of production is thus concealed by disguising the *discourse* (in which an enunciative source is present, its reference point is the present tense, and the pronouns "I" and "you" are engaged) in order to present itself as *history* [*story*] (in which the source of enunciation is suppressed, the verb tense is an indefinite past of already completed events, and the pronouns engaged are "he," "she," and "it"). Thus, in discourse the discursive *relation* is emphasized, while in history/story the address is impersonal.

For the spectator to have the impression that it is his/her own story being told, it must appear as if the fiction on the screen comes from nowhere. Since history is, by definition, "a story told from nowhere, told by nobody, but received by someone," the invisible style which hides the work of the enunciator makes it *seem* like "it is . . . the receiver (or rather the receptacle) who tells it."[19] The cinema can only function properly as an apparatus if the enunciating operations are hidden; in this way a "pseudo-viewer" is created, which every spectator can appropriate at will.

PSYCHOANALYSIS AND TELEVISION

I have explained the model for viewer participation in the cinema in such detail because the argument in psychoanalytic film theory is extremely complex; each interlocking part depends on its relation to the others. When we speak of unconscious desire in something as vast as the filmic institution, it is not simply a matter of applying psychoanalytic terms—a whole constellation of factors works together to produce what we call the film spectator. Given the theoretical situation and specific conditions in which the viewer in some sense "halluci-nates" the film, it should be clear that the paradigm for *television* view-ing must be radically different: a massive reformulation of the viewing process is required. In addition, compared to the extensive amount of work done in psychoanalytic film theory, there is relatively little work that approaches television from a psychoanalytic perspective.

The most obvious statement we can make is that the kind of "subject-effect"—or spectator—produced by the television apparatus will have to be quite different in a medium that *depends* on interruption for its mainspring. As John Ellis has pointed out, instead of demanding the sustained *gaze* of the cinema, TV merely requires that its viewer *glance* in its direction: "The gaze implies a concentration of the spectator's activity into that of looking, the glance implies that no extraordinary effort is being invested in the activity of looking. The very terms we habitually use to designate the person who watches TV or the cinema screen tend to indicate this difference. The cinema-looker is a spec-tator: caught by the projection yet separate from its illusion. The TV-looker is a viewer, casting a lazy eye over proceedings, keeping an eye on events, or, as the slightly archaic designation had it, 'looking in.'" [20] The TV viewer's attention is, at best, only partial (for all kinds of rea-sons, from the commercial "interruptions" to the domestic location of the TV set); there is a diffraction of the cinema's controlling gaze. For this reason, the TV viewer is not held captive in cinema's fascinating thrall.

From the very start, then, the spectator-dreamer analogy begins to crumble. Because there is no "artificial regression," primary voyeuristic identification is not engaged. The source of enunciation is dispersed (and made problematic), and with that, its terms of address. And, as we shall see, two of the most important features of the classical fiction film—the point-of-view and reverse-shot structures—are detached, partial, and isolated. A "fascination in fragments" is all that remains.

Since film theory's discussion of spectatorship is grounded, first of all, in the specific situation of viewing, we can be sure that different experiential factors will have profoundly different psychoanalytic con-sequences and effects. Films are seen in large, silent, darkened the-aters, where intense light beams are projected from behind toward luminous surfaces in front. There is an enforced and anonymous col-lectivity of the audience because, for any screening, all viewers are physically present at the same time in the relatively enclosed space of the theater. In contrast to this cocoon-like, enveloping situation is the fragmentary, dispersed, and varied nature of television reception. The darkness is dissolved, the anonymity removed. As Roland Barthes has pointed out (in considering television to be "the opposite experience" of cinema), the site of television reception is the home: "[T]he space is familiar, organized (by furniture and familiar objects), tamed. . . . Television condemns us to the Family, whose household utensil it has become."[21] While the aura of cinema spectatorship produces hypnotic fascination, the atmosphere of television viewing enables just the opposite—because the lights are more likely to be on, one can get up and return, do several things at once, watch casually, talk to other people, or even decide to turn the television off. In addition, the TV viewer can switch channels at will, enabling him or her to watch several shows simultaneously, and thus permitting the voluntary participation in any number of fictions at once. Robert Stam's suggestive metaphor appropriately compares the physical conditions of viewing: "It is not Plato's cave for an hour and a half, but a privatized electronic grotto, a miniature sound and light show to distract our attention from the pressure without or within."[22]

Closely related to the physical conditions of the viewing experience are the technological facts—the different technologies that generate cinema and television in turn produce specific psychoanalytic effects. A film is a strip of autonomous still images that appear to move when projected one after the other on the screen. One branch of film theory sees this as a major feature of the cinema's fascination: we perceive movement because we constantly see the alternation of presence (the image/frame) and absence (its divisions) when the strip of celluloid passes through the gate. The moving television image, on the other hand, is generated by the continuous scanning of whatever is in front of the camera by an electronic beam. An endless series of horizontal lines replaces the intermittent stillness of the single image.

Stephen Heath and Gillian Skirrow conclude that because of its electronic nature, the television image can be changed in the very moment of its transmission, creating a kind of "perpetual present" of the TV image which is very different from cinematic projection.[23] A film is always distanced from us spatially (we sit "away" from it in the theater), making the screen image seem inaccessible, beyond our reach. The television set occupies a space that is nearby—just across the room, at the end of the bed, or elsewhere. The television screen thus takes up a much-reduced part of the spectator's visual field, and seems available to us (the TV set is a controllable possession). It does not fascinate in quite the same way.

And it is precisely around TV's quality of "immediacy" that we can make some fundamental distinctions between psychoanalysis in film and television. For, as Heath and Skirrow point out, it hardly matters what content is communicated by the television, so long as the "communicating situation" is maintained. It is television's peculiar form of "presentness"—its implicit claim to be live—that founds the impression of immediacy: TV's electronically produced, present-tense image suggests a "permanently alive view on the world; the generalized fantasy of the television . . . image is exactly that it is *direct,* and direct for *me.*"[24] A sort of "present continuous" is created, confusing the immediate time of the image with the time of events shown, and thereby reducing the interplay of presence and absence that we find in the cinema (where we're always seeing the *present* image of an *absent* object/actor). Put another way, a film is always distanced from us in time (whatever we see on the screen has always already occurred at a time when we weren't there), whereas television, with its capacity to record and display images simultaneously with our viewing, offers a quality of presentness ("here and now") as distinct from the cinema's "there and then."

Although this argument is based on Heath and Skirrow's comments on live transmission and has been equally analyzed in Stam's work on television news and Jane Feuer's work on live TV, it is the *effect* of presentness—an illusory feeling—that I want to emphasize.[25] This impression of immediacy permeates everything from the more literally "live" programming of talk shows and televised news to soap operas and situation comedies. The only difference between the two limit-points noted here is that in the case of the former, an ideology of realness and spontaneity is emphasized, whereas in the latter it is only implied. John Ellis puts it this way: [T]elevision presents itself as an immediate presence. . . . Television pretends to actuality, to immediacy; the television image in many transmissions (news, current affairs, chat shows, announcements) behaves as though it were live and uses the techniques of direct address."[26] Whatever the format (though this is less so for prime-time serials which more closely resemble the cinema) television's "immediate presence" invokes the illusion of a reality presented directly and expressly for the viewer.

Thus television substitutes liveness and directness for the dream-state, immediacy and presentness for regression. It also modifies primary identification in ways that support its more casual forms of looking. The television viewer is a *distracted* viewer, one whose varied and intermittent attention calls for more complex and dispersed forms of identification. As we have seen, the cinema bases its primary identification on the association of the spectator's look with that of the camera. Television breaks down the voyeuristic structure of primary identification—there is no camera-position to be occupied in the same way. The "subject-effect" that results from primary identification in the cinema is thus fragmented, displaced, and multiplied into modes of suspension and delay. Primary identification in the cinema involves the spectator's identification with his own look and, by extension, with the camera; the film viewer is thus in a position to lend coherence/give meaning to the images. Television's fractured viewing situation explodes this coherent entity, offering in the place of the "transcendental subject" of cinematic viewing, numerous partial identifications, not with characters but with "views."

Remember that the psychoanalytic differences between film and television are rooted in technology. When we examine TV closely, we see that there can be a number of possible "looks," not of one camera, but of three. Three *different* types of camera looks will mean that the "constructed spectator" of television will be different from that of film. In discussing the different materials of the evening news, Stam outlines several different camera-looks (we can extend this logic to the types of looks found in any televisual "text"). There is the look of: 1) the film camera that shoots footage (this can be extended to those filmed episodes of prime-time dramas); 2) video cameras that tape material such as earlier reports (this can be extended to the taped material of soap operas, for example); and 3) tapeless video cameras directly transmitting images and sounds as in the anchors' situation (this can be extended to all "live" programming formats such as talk shows, magazine shows, and morning shows). Thus the television apparatus can be seen to generate, if not a multiplicity, then at least a variety of perspectives and camera positions with which to identify.

John Caughie discusses this dispersal not in terms of camera positions, but in terms of television's fragmented broadcast flow. Each little narrative unit, each little "drama" (and this includes such diverse elements as commercials, news briefs, and the like) provides the viewer with a sense of unity that is only momentary, to be disrupted again by the next TV segment. He concludes that there are numerous positions constantly available to the viewing subject, that coherence (as in identification) is dispersed, and that a notion of unity (as in the unified ego) can only be maintained through "recurrence," "repetition," and "routine structures."[27]

However, both Stam and Mimi White (in a recent article on television intertextuality) conclude that television does precisely the opposite. For Stam, this variety actually enhances the viewer's ability to be everywhere at once, "granting an exhilarating sense of visual power to its virtually 'all-perceiving' spectator, stretched to the limit in the pure act of watching."[28] White maintains that television's diverse interprogram references create a totalizing world that binds the separate fictions and the reality of actual events into one continuous whole. She maintains that the interruptive variety of television does in fact produce a unified subject across individual programs, days, and so on, in the production of a world that is "progressively all-encompassing, self-defining and continuous." She concludes that, in fact, American commercial television does "engage in practices that assert unity and address the spectator-as-ideal-subject across temporal, spatial, and narrative diversity.[29]

Yet it is not simply the unifying effect, the imaginary coherence, that links primary identification to the apparatus. As we have seen, in the cinema the processes of identification are connected to authorship, and the work of the filmmaker is conceived in terms of enunciation: "I [the subject] espouse the filmmaker's look (without which no cinema would be possible)."[30] Such a concept of authorship is literally nonexistent in television, where the practical implications of programming make such centrality impossible. Who is the author of *The Love Boat*, or of *Jeopardy*, or of the *CBS Evening News* for that matter? The television apparatus makes us redefine the notion of "author"—and with it the "enunciative source" in television. How can we speak of this new "producer of the fiction" in television? What evidence can we find of unconscious desire, and how does *this* "enunciator" address us differently than the one in film?

You will remember that in the theory of cinematic enunciation, each filmmaker possesses and then delegates *the look,* sometimes through fictional substitutes (characters) or at times intervening at certain privileged moments. This is what characterizes a particular director's system of enunciation—the organization of a system of looks across the viewer's visual field. In television, by contrast, the look is much more qualified and diffuse; partial views are almost never negotiated through a character in most television forms—game shows, talk shows, television news, to name but a few. But even in its closest approximation, the prime-time serials, this system of looking is complicated. If the enunciative source is conceived of as a "desiring producer" or a site of unconscious desire, how is this defined in *Dynasty,* for example? Is the authorial subject-position held by Richard and Esther Shapiro, by Aaron Spelling, by E. Duke Vincent, by the individual director of an episode, or by the sponsor? It is only possible to say, then, that the look that hovers over the television text is disembodied and dispersed. Because we only sometimes see through the eyes of a character (proportionally speaking, the bulk of TV programming gets along quite well without these constructions), our look is most often *not* mobilized in a fictive exchange. Heath and Skirrow suggest that television constructs "the fiction of a position outside of any system, the construction of the position of a *look on* as unimplicated in address."[31] The "who speaks" (or "whose desire is articulated") of cinematic enunciation becomes the position of the spectator as a look.

This issue might be made a little clearer if we look at some recent remarks by French filmmaker Jean-Luc Godard. In an hour-long video entitled *Soft and Hard* (1985), Godard and his collaborator Anne-Marie Mieville discuss their work, the differences between film and television, the language of visual images, and so forth. This is all juxtaposed with clips from classical Hollywood films and contemporary TV programming. At one point near the end, Godard speaks of his work as a filmmaker, and of his frustrations in television. When seen in the light of film theory's argument about enunciation—and the changes called for by the different enunciative structure of TV—these comments seem remarkably astute. Enunciation is a concept *linked* to authorship, but it is not exactly the same thing. The notion of the unconscious as a productive source is what marks the difference between the two.

> When one says "I," you can see that . . . "I" projecting itself towards others, towards the world. . . . The cinema has shown that quite clearly, more than all other forms. . . . The "I" could be projected, enlarged, and could get lost. But its idea could be traced

back. Television, on the other hand, can project nothing but *us* [elle *nous* projette], so you no longer know where the subject is. In cinema, in the very idea of the large screen, like in the myth of Plato's cave, [we have] the idea of "project," "projection," which in French, at least, have the same roots. Project, projection, subject. With TV, on the other hand, you feel that you take it in [on la reçoit]/you're subjugated by it, so to speak. You become its subject . . . like the subject of a king.[32]

Another difference between film and television that is partly related to this issue of enunciation has to do with the television "text." It involves both the form and content of the "classical cinema"—whose aim is the construction of a fictional world in which the illusion of reality is provided by the fluid continuity of seamless editing. In a very basic sense, there is nothing that corresponds to the feature-length film in television. Even the miniseries and the made-for-TV movie (which are organized pretty much along the lines of the dominant fiction film) are marked by the segmentation and variety characteristic of television. At the same time, the serial form of soap operas and prime-time dramas implies that we will always be frustrated in our desire for narrative closure. In television, our need for such completion becomes reorganized; elements of the story are partially resolved in a way that inevitably permits the continuation of the text. For example, where classical Hollywood cinema almost always leads to marriage (the formation of the couple being seen as synonymous with resolution), marriage in most television narrative forms (especially in soap operas and prime-time dramas) is a major mode of complication, a site of disruption rather than resolution. The unstable, reversible, and circular movement of this type of program thus frustrates our desire for closure, for it embeds interruption into the very heart of the discursive structure. Furthermore, instead of classical cinema's process of "narrativization" (described by Stephen Heath as a movement in which the organization of space and time is subsumed by the causal relations of narrative),[33] television provides a fragmentary and discrete series of microstagings. Therefore, even in TV's most fictional forms—those places to which we would most readily look for the similarities between television and film—the TV apparatus organizes spectatorship quite differently.

Finally, one of the most important differences between film and television, when analyzed in terms of psychoanalysis, involves the point-of-view and reverse-shot structures. In the cinema, these editing figures both have to do with the spectator's ability to construct an *imaginary coherence*—a filmic space-time dimension—from film's peculiar ability to order spatial and temporal relations. In the classical model of the fiction film, properties of narrative storytelling, seamless editing, and secondary identifications with characters contribute to the production of an *illusory world* with its own internal consistency, which may or may not correspond to a "real" referent outside the text. Historically, it was through editing, the joining of shot to shot in the construction of this fictional world, that the cinema came to have its own method of constructing not only "reality" but its spectator as well. However, as we have seen from the theory of enunciation, the work of the organizing principle (the "author")—a subject external to the text which selected and arranged the shots into a composite fictional world—had to be rendered invisible. The rules of continuity were developed in order to maintain the impression of an imaginary coherence, permitting the spectator's belief in the integrity of the space, the logical sequence of the time, and the "reality" of the fictive universe.

Most often, the spectator's ability to construct a mentally continuous time and space out of fragmentary images is based on a "suturing" system of looks, a structured relay of glances: 1) from the filmmaker/enunciator/camera toward the profilmic event (the scene observed by the camera); 2) between the characters within the fiction; and 3) across the visual field from spectator to screen—glances that bind the viewer in a position of meaning, belief, and power. It is these traversing gazes that are primarily negotiated through the reverse-shot and point-of-view structures, the central means by which "the look" is inscribed in the cinematic fiction. Most commonly applied to conversation situations, the reverse-shot structure implies an alternation of images between seeing and seen, the point-of-view shot anchoring the image in the vision and perspective of one or another character. The spectator therefore identifies, in effect, with someone who is always off-screen, an absent "other" whose main function is to signify a space to be occupied. In psychoanalytic terms, the spectator is inserted into a logic of viewer/viewed which evokes certain unconscious fantasy structures such as the primal scene (an early "scenario of vision" in which the unseen child observes the parents' lovemaking). Film theory suggests that it is just such a combination of vision and desire which lays the groundwork for a comparison between film viewing and unconscious activity.

Therefore, in the cinema, the reverse-shot structure enables the spectator to become a sort of invisible mediator between an interplay of looks, a fictive participant in the fantasy of the film. From a shot of one character *looking,* to another character *looked at,* the viewer's subjectivity is bound into the text. However, this positioning of the spectator as a sort of ideal voyeur is totally broken down in television. In her discussion of live TV, Jane Feuer concludes that the form of reverse shot prevalent in the morning show, for example, denies the expected responding shot: "Instead of placing us within the world of the show, the fragmentation of space actually places the viewer *outside* the system." However, Feuer admits to having difficulty theorizing the *level* at which this positioning occurs, opting for the somewhat vague description of a "netherworld between 'effects' and 'unconscious work.'"[34] What I am suggesting here is that, where in the cinema the reverse-shot structure works together with the point-of-view system to bind the spectator into a position of coherence and fictive participation, in television, the effect is just the opposite.

As we shall see, even the reverse-shot structure, the staple of soap opera (a paradigmatic televisual form)—whose continual exchange of dialogue often provides the only basis for the drama—is drastically changed in television because of the fragmentation and dispersion I've already discussed.[35] Where in the cinema this binding operation has been used to perpetuate what Noël Burch has called the cinema's "greatest secret"—in which the fragmented space is recombined through editing to preserve an illusory, fictive continuity—television needs no recourse to such disguise.[36] In the soap opera, it is never a question of "creating" a coherent space, concealing the activity of an organizing principle outside the text, because the spectator is never *moved* through space. A belief in the fictional totality is not necessary, for what the reverse-shot accomplishes in the soap opera is something altogether different. The quality of viewer involvement, instead, is one of continual, momentary, and constant visual repositioning, in keeping with television's characteristic "glance." The look is not focalized, as it is in a classical film by a director such as Alfred Hitchcock or Fritz Lang, for example; there is no enunciator to transform the discourse into an apparently self-generating story.

GENERAL HOSPITAL: A CRISIS IN REPRESENTATION

We can get some idea of the vast differences between film and television through a close analysis of a specific soap opera episode—one broadcast on *General Hospital* in February 1986. The episode is both characteristic and atypical of the soap opera genre. It is atypical of the soap opera's relatively confined format because it contains a large number of outdoor "remote" segments, filmed on location in Charleston, South Carolina. But in its composition (its basic patterns of enunciation, its systems of looking, its terms of address) the episode is, in fact, quite representative of the way that daytime dramas produce their meanings and engage identification on the part of viewers.

Within the space of a program hour (minus the fifteen minutes for commercials) we can find an incredible variety and complexity of both shot setups and narrative subsegments. Our vision is thus dispersed, fragmentary, and amplified; this quality of viewing is both characteristic of the soap opera form, and central to the peculiar kind of spectatorship in television I have been describing. Roughly, the day's narrative involves the wedding of Terry Brock and Kevin O'Connor, which takes place in their hometown of Laurelton. Two clouds hang over this marriage—an undefined and unnameable crisis of three years earlier involving Terry and the whole town that is deeply repressed in Terry's psyche, and two unsolved murders of Laurelton men that took place in Port Charles, the regular setting of *General Hospital*. While all of Terry and Kevin's close friends from Port Charles are at the wedding, police chief Anna Devane is continuing her murder investigation. In addition to Bobbi Brock (Terry's stepmother, best friend, and O. R. nurse at General Hospital) and Jake Myer (a lawyer and Bobbi's boyfriend), Frisco Jones (a rock singer turned police cadet) and Felicia Cummings (Frisco's fiancée and aspiring businesswoman), Tony (Frisco's brother and a doctor at General Hospital) and Tanya (his wife, a speech therapist), and Kevin's brother Patrick (an intern)—who all live in the brownstone where the murders took place—Jennifer Talbot (Terry's grandmother, a Laurelton matriarch who is controller of the town and guardian of its secret), Ted Holmes (Jennifer's lawyer and associate), and the O'Connor parents are all present at the wedding. Within the course of the episode, Anna learns that she has reason to suspect either of the O'Connor brothers as murderers and rushes to Laurelton to try and prevent the wedding from happening.

This complexity of characters and narrative content is matched by an equally complicated structural form. The episode can be broken down into roughly sixty-seven distinct narrative segments (this division suggests a number of analytic problems that will be discussed shortly). There are close to six hundred individual shots in the episode, with most sequences averaging around ten. Thus we can find an astounding variety of shot setups within the confined world of the soap opera; although there is a limited number of locations for each scene (ordinarily the hospital, the brownstone, other apartments, the waterfront cafe, etc., but in this case the Talbot living room, the bedroom, the street, the church, etc.) there is a great diversity of camera angles and distances. Camera distance is further complicated by a continually moving (or zooming) camera that often rests momentarily on a conversation before moving again. For this reason, there is a perpetual "fracturing" of the televisual space. For example, we might see the repeated transition from a medium-shot of a group with considerable background activity, to individual close-ups. And these are almost never simply a repetition of a single glance, but involve a constant diversion of the eyes or a reframing of the space (to include a portion of another character, a different angle, etc.). Changes of shot are triggered by dialogue and there is relatively little sound overlap; as each character speaks, the camera is on her or him. From this, a visual rhythm which depends on fragmentation is built. Within a single sequence there is never any *sustained* camera work (the camera hops about from place to place), nor any sustained focus of the representation. Rather, we find a parallel on the formal level (at the level of the visual signifier) to what occurs in the drama (at the level of the narrative signified); all these complexities make television viewing the kind of experience that is never still. The soap opera form confirms that TV viewing means the eye is constantly in movement; it never rests, and always has something new to see.

This constant motion clearly implies that there is no "unifying presence" at the site of spectatorship, and as such goes to demonstrate just how powerfully television reorganizes our patterns of looking. Soap opera, as an exemplary television text, mobilizes different relations of desire and vision than those that operate in the classical narrative fiction film. Even when it uses the structures of classical cinema (point of view and reverse shot, the flashback, systems of continuity and alternation, a focus on the image over dialogue/sound, and the special attention to the figure of the woman-as-display), the daytime drama has to modify these, reworking them to fit in with the televisual system of enunciation. This is what I mean by a "crisis in representation": each of classical cinema's founding structures has to be modified because of the inability of soap opera's televisual form to comfortably reproduce cinematic structures of meaning. As we shall see, at certain crucial moments when television (the soap opera) "tries to be cinema," something happens that radically disturbs the TV text.

The first point to be made concerns the construction of an imaginary space. As I noted, the cinema spectator's fictive participation hinges on a perceived spatial coherence—fragmentary images are given a logical consistence because they are subordinated to a causal sequence of narrative events. Stephen Heath refers to this process as "the conversion of seen into scene," in which vision itself is dramatized, staged as a narrated spectacle before the viewer.[37] However, the soap opera's form disturbs such a narrative binding, giving us not a dramatic "scene," but an infinite variety of autonomous "seens," partial views of interrupted exchanges. Since the concern of the soap opera is not to build coherent spatial relationships but to focus on isolated moments of interaction, the masking operation of classical cinema is necessarily absent here. Because of the way space is continually dispersed and reorganized, and because each scene is repeatedly and insistently fragmented, we don't find the same illusory space-construction that was such a central part of the film viewer's fictional role.

Space is continually *redefined* by a camera movement without cuts, and individual shots are then related by background elements that unify the space. Since the profilmic space of the soap opera is already confined and coherent (one could say that there is a *preexistent* unity), an artificial space is not constructed by the viewer in the same sense. Instead, there is a subtle (but inevitable) variation that comes about through secondary characters, backgrounds, objects, and different visual configurations. Because this variation often occurs within a single shot (and not from the movement across the cut), the principles of continuity editing are not used to disguise the transitions between shots. Therefore, rather than the continuous move to incorporate space and suppress difference, the "seen" of the soap opera emphasizes diversity; what the viewer sees is a perpetual scene-in-flux.

Unity is provided by another source. The typical sequence form of *General Hospital* is what Metz describes as the cinematic "scene."[38] In this particular form of organizing temporal and spatial relations (Metz says that it is the most theatrical), a unified, continuous time and space are represented. Whereas the "sequence" for Metz is the most cinematic (because it depends on the spectator's ability to skip over the gaps—those short ellipses in time and space which Burch describes as the "secret"), the "scene" involves a fairly static and unified *preconstructed* space. Furthermore, the difference between spatial representation in film and TV is due, in part, to the fact that the soap opera "scene" is generally profilmicly continuous, being recorded by multiple cameras simultaneously, whereas the "single camera" style in cinema necessarily involves a fragmented profilmic event. In TV, the diversity comes instead from the variations within a single shot and between the different shots. It is not necessary for the spectator to construct a space because the depicted scene is already coherent and whole. Thus a continuous and uninterrupted signified (a conversation, for example) is matched by a fragmentary signifier. When the drama is organized along these lines, then, the way the spectator becomes involved is by means of minute gaps, divisions, and modifications within a continuous whole. It should be easy to see that the soap opera's format of limited attention is especially suited to this type of visual style.

We can see this spatial variation in three early sequences of the episode, where reference to Terry's troubled past is made when she says she wants to walk triumphantly down Laurelton's main street in her wedding gown. This sets into place the fantasm or memory of the past that haunts the entire episode and gives it its dramatic core. Because of the spatial play that I've described, the soap opera text cannot be segmented or broken down in the way that a classical film text can. Each conversation—between Terry and her grandmother Jennifer in the bedroom, between Jennifer and Kevin in the hallway, and finally between Terry and Kevin on either side of the door—is linked by a shot that forms a part of both the segment that precedes it and the one that follows. This *chain-type* of organization creates the impression of the complete contiguity of space (though the conversation moves from room to room) and continuity of time (maintained by the dialogue). Yet, at the same time that we have this spatial and temporal smoothness, there is an endless variety in *types* of shots (from medium-long-shot disclosing the entire bedroom, to medium-shot, and to the range of closer shots).

But more important than this, and crucial to the soap opera's reorganization of the "look," is the peculiar variation of the reverse-shot structure that we find in even the most banal eyeline "exchange." For rather than a systematic volley of alternated looks, here we have a close-up of one character invariably exchanged with a more distant shot of the other (which locates him or her in a background filled with objects). Furthermore, the soap opera viewer is never given a responding shot that would indicate the *perceptual* point of view of the person speaking. Thus close-ups often appear from nowhere, or seem to, without the spatial anchoring we find in classic film. In addition, the eyeline is continually diverted, such that the look is never directly at the other character, but always somehow curiously askance. This creates a *dislocation* of the glance that utterly negates the conception of off-screen space that is so central to the cinema. For in film, off-screen space exists as a location to be filled and naturalized—an off-screen absence becomes present in the following image. From the displaced sidewise glance that we have here, then, the cornerstone of spectator-positioning in the cinema begins to collapse.

This is exactly the pattern established in the conversation between Terry and her grandmother. However, this displaced eyeline structure which characterizes the daytime drama is even more distorted in the following conversation between Terry and Kevin. Not accidentally, the curious setup of this conversation is necessary because the body of the woman must not be seen, Jennifer says to Terry: "You can't let him see you, it'll be bad luck." (This prohibition of looking at the woman's body, which—we'll see later—structures the entire episode, has already been referred to in a conversation between Terry and her grandmother. When Terry says she needs to confront the past, Jennifer responds, "No, not like this, not by making a spectacle of yourself.") Kevin and Terry are thus on opposite sides of a door that separates them, while the conversation is depicted in an alternation between two types of shot. One shot shows the two of them pressed against either side of the door, the other features close-ups of Terry. The peculiar way we address someone when we can hear but cannot see them thus becomes the basis for this visual exchange. In a striking diversion of the eyeline match—which also ingeniously avoids the need for cutting (by containing the two speakers within a single image)—the spectator is given a view, not of two characters looking, but of two characters precisely *not* looking. This is an emphatic example of the "disturbance of glances" that characterizes the soap opera as a model of televised form.

The traditional point-of-view shot, a mainstay of cinematic identification, is disturbed in another way by the soap opera structure. Conventionally, the point-of-view shot, often in combination with the subjective image, is one of the primary ways of inscribing the spectator's subjectivity. It anchors cinematic identification by making the spectator's glance coincide with that of a specific character. As we have seen by the lack of even the simple perceptual point-of-view shot in the conversational situation, it is highly unusual for there to be such an orchestration of subjectivity and vision in the soap opera. In this episode, however, critical moments of deep disturbance—which link memory, transgression, and sexuality for the woman—are signalled by the use of such point-of-view shots. Yet, importantly, the text can neither master nor control these shots, forcing them to remain as moments of rupture in a striking example of the soap opera's difficulty with the classical cinema's system of vision.

Three moments in the episode are marked by instances of Terry's point of view. The first sets up the point-of-view figure and prepares the way for the following two sequences. The second signals the beginning of the traumatic flashback, and the third points to elements of a future significance in the serial. In the first, Terry and Jake (who will give her away) are riding in a ceremonial old-fashioned car on their way to the church. They stop to remark on the steeple of another church, shown from a low angle that approximates the site of their vision. However, the corresponding shots of them—shown from above as they look upward—simply fragment the space rather than locating them (and the spectator) as points of coherence which anchor the vision. In the second segment, there is an indication of something strange, as haunting music plays over Terry's horrified and confused gaze, first at a motel, and then at the wedding church. However, this five-shot alternation between seeing and seen is cut off abruptly by the commercial break, making it impossible to fully integrate the sequence into any sustained system of vision. Finally, after Terry has arrived at the church, she abruptly stares again (the haunting music returns)—

this time toward the town hall. The purpose of this five-shot exchange between a close-up of Terry and the hall (which goes in and out of focus) is to make the place seem mysterious and troubling. Again, the use of a camera device (the manipulation of focus) is an isolated figure; we have not seen it before nor will we see it again. The figure of focus-distortion is thus never *integrated* into a system which emphasizes the act of seeing. Each of these moments, then, asserts a kind of "textual difference" from the rest of the soap opera episode. Because of this, they frustrate any possibility of anchoring spectator-identification within the subjective vision of the character.

What the look at the motel triggers, in fact, is the traumatic core of both the individual day's episode and the extended text of *General Hospital* itself. The sight of the motel sign activates a memory of its neon glow, which in turn brings on a flashback of the night when Terry—dazed, hysterical, and naked—walked down the main street of Laurelton, half sobbing, half praying. A thirteen-shot sequence follows, involving two types of shot: The first type depicts Terry in close-up, camera at an oblique and unsettling low angle, tracking her as she walks, while the second shows people on the street watching her go by. This sequence is marked as a point-of-view sequence, yet the soap opera's system of looking cannot support the traditional structure. Instead, there is a series of bizarre, unmatched angles, a subjectivity not narrativized in the conventional cinematic way. Crucially, the low-angle shots of Terry cannot possibly be construed as the point of view of the onlookers; the gaze at Terry thus resists anchoring in these fictional characters. Likewise, the shots of the onlookers are in no way portrayed from Terry's point of view, for they look back toward her after she has passed. There are no "fictive alibis" to make the gaze seem plausible. Instead, what becomes foregrounded is this—the very act of watching the woman as spectacle in a virtual performance of sexual transgression and excess.

The flashback here is a memory, not only for the fiction and its characters (whose personal experience it represents), but of a form of meaning-production—the cinema/point of view—that haunts the dispersed and fragmentary structure of the soap opera. For in the endless text of *General Hospital,* the memory makes its return, appearing in future episodes through music, dialogue, or images. All that remains of this frustrated attempt to organize the point of view are images and sounds, moments of fragmentary fascination to be rearticulated in the text. It is no accident that this crisis of identification should be expressed through the body of the woman. Classical cinema can be said to repeatedly organize its fictions around an image of the woman, object of desire in endless pursuit. The soap opera form enjoys no such centrality, no such coherence and integration. That this critical memory sequence is the nightmare of a woman on display reminds us of the traces of the classical cinema that appear throughout the soap opera text. But Terry is not a fictional substitute for the viewer, either; our gaze neither remains with her nor evokes a feeling of empathy for long. The creation of a momentary *experience* in which time, memory, and fantasy converge, is rather what the sequence gives us, in keeping with television's fragmented subjectivity and its dispersion of viewing.

One final point about this episode concerns Terry's triumphal wedding march down that same street (after the flashback memory), as she purposefully sings a hymn: "I am so happy, His way is my way." Yet at the very moment that she is ridding herself of her past, Anna Devane is trying desperately to stop the wedding. What happens at this point, then, is a form of alternation very common in the cinema but rare for the daytime television text. Sequences of Anna running through the town alternate with those of Terry singing and walking toward the church. While crosscutting (the simultaneous representation of action in disparate spaces) is the cornerstone of cinematic storytelling, what marks this series of shots as distinctly uncinematic is the absence of sustained continuity editing. There is little effort to match action in such a way as to produce a continuous and transparent flow. Rather than the effort to portray simultaneity, the emphasis here, instead, is on the individual action in itself, for *General Hospital*'s enunciative system, based as it is on fragmentation, has little use for cinematic transparency.

From these examples it should be clear that the classical cinema's unified form contrasts the fragmentation at the heart of the televised daytime drama. Both film and television have specific systems of enunciation that structure relations of vision and identification in different ways. These terms of address each produce a different type of spectator, a different subject-effect for each mode of meaning-production. I have tried to show by this example how virtually every psychoanalytic process in the cinematic apparatus is deconstructed by the complex strategies of enunciation in the soap opera (as a prime example of the TV apparatus). Yet "deconstruction" is not the precise term, for, rather than being a subversion, it seems instead that the systems of fascination and visual pleasure, the erotic engagement of processes of looking, are merely organized *differently* in the television text.

CONCLUSION

I have tried, in this analysis, to show how the cinematic apparatus is a machine of fascination, luring the spectator into desire for the image. The apparently innocent act of cinema-viewing involves unconscious factors of which we may not even be aware, for it engages multiple processes of the psyche in its task. The television apparatus is equally fascinating, yet it provides a very different kind of lure. Blurring the categories of fiction and nonfiction, embedding distraction in its very core, instilling a desire for continual consumption (not only of its programs but of the products that it sells), trading on the powerful sense of immediacy that it creates, the television apparatus is in many ways more pervasive than its kin. Both the cinema and television are combined technological and libidinal institutions, creating spectators insistent on perpetual return. Yet in television, a complex network of ratings, consumption, and economic exchange requires ever more powerful psychic mechanisms, reduplicating structures of fascination to compensate for its appeal to a dispersed and fractured subjectivity. The very nature and function of our fantasmatic participation in the televisual situation must be redefined. Early in this essay I cited Christian Metz's formulation of the cinematic institution as a form of "mental machinery" that has adapted spectators to the consumption of films. He sees its function as the production of pleasure, for "the cinema is attended out of desire, not reluctance." With a few modifications, this description could apply to television as well. And yet psychoanalysis, which provides a way of understanding how the cinema operates, can only provide us with a series of questions where television is concerned. The mechanisms that produce and regulate desire in television are infinitely varied, multitudinous, and complex. The field is open; while an elaborate psychoanalytic model of film spectatorship exists, the work in this area of television remains to be done.

By tracing out the terms of psychoanalysis in film studies, and by offering a suggestive example of television's differences, I've hoped to indicate some of the things we need to think about in developing our own theories of spectatorship in TV. For as Metz points out, all of us— analyst, critic, and spectator alike—are fueled by the workings of unconscious desire.

NOTES

1. This discussion relies in part on Terry Eagleton's very useful summary discussion of psychoanalysis in *Literary Theory: An Introduction* (Minneapolis: University of Minnesota Press, 1983), pp. 151–93.

2. Christian Metz, "The Imaginary Signifier," *Screen* 16, no. 2 (Summer 1975): 14–76. This article also appears in a book of collected essays by Metz, *The Imaginary Signifier: Psychoanalysis and the Cinema* (Bloomington: Indiana University Press, 1982), pp. 3–87. The quote is from p. 17 of the book; all subsequent references will be to this book.

3. Ibid., p. 7.

4. Jean Laplanche and Jean-Baptiste Pontalis, *The Language of Psycho-Analysis* (New York: W. W. Norton, 1974), p. 475.

5. Both essays are found in Theresa Hak Kyung Cha, ed., *Apparatus* (New York: Tanam Press, 1980), pp. 25–37 and 41–62.

6. Octave Mannoni, *Clefs pour l'Imaginaire* (Paris: Eds du Seuil, 1969), pp. 9–33.

7. Baudry, "The Apparatus," p. 56.

8. Ibid., p. 58.

9. Bertrand Augst, introduction to *Christian Metz: A Reader* (Berkeley: n.p., 1981), p. 3.

10. Metz, "Imaginary Signifier," pp. 48–49.

11. Ibid., p. 49.

12. Baudry, "Ideological Effects of the Basic Cinematographic Apparatus," p. 34.

13. Sigmund Freud, "The Poet's Relation to Day-Dreaming," in *On Creativity and the Unconscious* (New York: Harper and Row, 1958), p. 51.

14. Emile Benveniste, "L'appareil formel de l'énonciation," in *Problèmes de linguistique générale* (Paris: Eds du Seuil, 1974), p. 85.

15. Raymond Bellour, "Alternation, Segmentation, Hypnosis: Interview with Raymond Bellour by Janet Bergstrom," *Camera Obscura* 3/4 (Summer 1979): 98.

16. This article was originally published in *Edinburgh 76 Magazine/Psychoanalysis and the Cinema* #1 (1976): 21–25, which is the text I am citing. It

also appears in Metz, *Imaginary Signifier*, under the title "Story/Discourse: A Note on Two Kinds of Voyeurism." The original version can also be found in John Caughie, ed., *Theories of Authorship* (London: Routledge and Kegan Paul, 1981).

17. Metz, "History/Discourse," p. 24 (italics mine).

18. Bertrand Augst, "The Order of [Cinematographic] Discourse," *Discourse #1* (1979): 51.

19. Metz, "History/Discourse," p. 24.

20. John Ellis, *Visible Fictions: Cinema, Television, Video* (London: Routledge and Kegan Paul, 1982), p. 137.

21. Roland Barthes, "Upon Leaving the Movie Theater," in *Apparatus*, p. 2.

22. Robert Stam, "Television News and Its Spectator," in *Regarding Television—Critical Approaches: An Anthology*, ed. E. Ann Kaplan, American Film Institute Monograph Series, vol. 2 (Frederick, Md.: University Publications of America, 1983), p. 27.

23. Stephen Heath and Gillian Skirrow, "Television: A World in Action," *Screen* 18, no. 2 (Summer 1977): 54.

24. Ibid.

25. Jane Feuer, "The Concept of Live TV," in *Regarding Television*, pp. 12–22.

26. Ellis, *Visible Fictions*, p. 106.

27. John Caughie, "The 'World' of Television," *Edinburgh 77 Magazine #2: History/Production/Memory* (1977): 81.

28. Stam, "Television News," p. 24.

29. Mimi White, "Crossing Wavelengths: The Diegetic and Referential Imaginary of American Commercial Television," *Cinema Journal* 25, no. 2 (Winter 1986): 62.

30. Metz, "Imaginary Signifier," p. 55.

31. Heath and Skirrow, "Television," p. 46 (italics mine).

32. Thanks to Janet Perlberg and Karen Cooper of the Film Forum in New York City for making the tape available to me.

33. Stephen Heath, "Narrative Space," in *Questions of Cinema* (New York: Macmillan, 1981), p. 43.

34. Feuer, "Concept of Live TV," pp. 18, 20.

35. After I had completed a version of this essay, Jeremy Butler's "Notes on the Soap Opera Apparatus: Televisual Style and *As the World Turns*" appeared in *Cinema Journal* 25, no. 3 (Spring 1986): 53–70. I was struck by the similarity in both our approaches and some of our conclusions (which we had arrived at independently) and suggest Butler's article as an extremely interesting and useful supplement to this chapter.

36. Noël Burch, "Film's Institutional Mode of Representation and the Soviet Response," *October* 11 (Winter 1979): 82.

37. Heath, "Narrative Space," p. 37.

38. Metz, "Problems of Denotation in the Fiction Film," in *Film Language: A Semiotics of the Cinema* (New York: Oxford University Press, 1974), pp. 108–46.

FOR FURTHER READING

There are a number of primary texts of psychoanalytic film theory; all are central to formulating key concepts in the relationship between psychoanalysis and the cinema (apparatus, gaze, identification, split belief, mirror stage, etc.).

Christian Metz, *The Imaginary Signifier: Psychoanalysis and the Cinema* (Bloomington: Indiana University Press, 1982) contains the following essays by Metz: "The Imaginary Signifier," pp. 3–87; "The Fiction Film and Its Spectator," pp. 101–47; "History/Discourse: Notes on Two Kinds of Voyeurism," pp. 91–98; "Metaphor/Metonymy," pp. 151–314. "The Imaginary Signifier," perhaps the most comprehensive "statement of purpose" in the field, discusses the unconscious structures that underlie our experience of film, noting how the powerful impression of reality in cinema is first and foremost an illusion. "The Fiction Film and Its Spectator" explores the analogies and disanalogies between film and dream. All the essays use concepts from Lacanian psychoanalysis, but "The Imaginary Signifier" contains the most complete application.

The anthology *Apparatus*, edited by performance artist Theresa Hak Kyung Cha (New York: Tanam Press, 1980), contains several of the founding articles of psychoanalytic film theory. Many of the essays deal with the spectator's experience as a semihypnotic trance; with the similarities and differences between film and dream; and with the notions of regression, identification, and the cinematic apparatus. All are important, but among the most significant are: Bertrand Augst, "The Lure of Psychoanalysis in Film Theory," pp. 415–37; Roland Barthes, "Upon Leaving the Movie Theater," pp. 1–4; Jean-Louis Baudry, "Ideological Effects of the Basic Cinematographic Apparatus," pp. 25–37, and "The Apparatus: Metapsychological Approaches to the Impression of Reality," pp. 41–62; Thierry Kuntzel, "The Defilement: A View in Close-Up," pp. 233–47. Metz's "The Fiction Film and Its Spectator" is also included in this anthology, pp. 373–409. In addition, highly interesting articles by filmmakers Dziga Vertov, Maya Deren, Jean-Marie Straub, and Daniele Huillet and a conceptual piece by Cha are included.

Janet Bergstrom's interview with cine-semiologist Raymond Bellour, "Alternation, Segmentation, Hypnosis," *Camera Obscura* 3/4 (Summer 1979): 70–103, is a useful, explanatory article describing key concepts in a conversational tone. Stephen Heath's collection of essays, *Questions of Cinema* (New York: Macmillan, 1981), is more difficult reading but combines important psychoanalytic generalizations with textual analysis; see especially the essay entitled "Narrative Space," pp. 19–75. And Laura Mulvey's landmark article, "Visual Pleasure and Narrative Cinema," *Screen* 16, no. 3 (Autumn 1975): 6–18 (and reprinted in several anthologies), discusses the psychoanalysis of spectatorship in terms of sexual difference.

There are a number of critical overview articles that summarize the issues in psychoanalytic film theory. The most useful collection, though often extremely difficult reading, is one put out by the British Film Institute for the Psychoanalysis and Cinema Event in Edinburgh in 1976, *Edinburgh 76 Maga-*

zine #1: Psychoanalysis and Cinema (London: British Film Institute, 1976). Among the articles included are Rosalind Coward, "Language and Signification: An Introduction," pp. 6–20; Stephen Heath, "Screen Images, Film Memory," pp. 33–42; Claire Johnston, "Toward a Feminist Film Practice: Some Theses," pp. 50–57; Christian Metz, "History/Discourse: A Note on Two Voyeurisms," pp. 21–25; and Geoffrey Nowell-Smith, "A Note on History/Discourse," pp. 26–32.

Janet Bergstrom, "Enunciation and Sexual Difference," *Camera Obscura* 3/4 (Summer 1979): 32–69, is another useful discussion of important concepts. For a more critical (and sometimes skeptical) perspective on psychoanalysis, see Dudley Andrew's chapter entitled "Identification" in *Concepts in Film Theory* (New York: Oxford University Press, 1984), pp. 133–56; Charles Altman, "Psychoanalysis and Cinema: The Imaginary Discourse," *Quarterly Review of Film Studies* 2, no. 3 (1977): 257–72; and Christine Gledhill, "Developments in Feminist Film Criticism," in *Re-Vision: Essays in Feminist Criticism*, ed. Mary Ann Doane, Patricia Mellencamp, and Linda Williams (Los Angeles: American Film Institute, 1984), pp. 18–48.

Individual articles that describe the psychoanalytic method or concentrate on particular textual analyses can be useful in clarifying the major points. Some of these are: Raymond Bellour, "Hitchcock: The Enunciation," *Camera Obscura* 2 (Fall 1977): 66–91, and Sandy Flitterman, "Woman, Desire, and the Look: Feminism and the Enunciative Apparatus of Cinema," in *Theories of Authorship*, ed. John Caughie (London: Routledge and Kegan Paul, 1981), pp. 242–50, which both discuss *Marnie*; two articles by Thierry Kuntzel that discuss *The Most Dangerous Game* ("The Film-Work, 2," *Camera Obscura* 5 [Spring 1980]: 6–69, and "Sight, Insight, and Power: Allegory of a Cave," *Camera Obscura* 6 [Fall 1980]: 90–110); and Stephen Heath's detailed analysis of *Touch of Evil* ("Film and System, Terms of an Analysis, Part II," *Screen* 16, no. 2 [Summer 1975]: 91–113). In "The Order of [Cinematographic] Discourse," *Discourse #1* (1979): 39–57, Bertrand Augst applies Foucault to discussions of the cinema, and in "That 'Once-Upon-a-Time . . .' of Childish Dreams," *Cine-tracts* 13 (Spring 1981): 14–26, Sandy Flitterman discusses Freud's theory of "family romance" in relation to the cinema. Mary Ann Doane discusses identification in "Misrecognition and Identity," *Cine-tracts* 11 (Fall 1980): 25–32; and Lesley Stern discusses point of view in "Point of View: The Blind Spot," *Film Reader*, no. 4 (1979): 214–36.

As my chapter indicates, there is relatively little already written in the field of psychoanalytic television studies. Still, there are a number of articles that take important steps in that direction, and John Ellis, *Visible Fictions: Cinema, Television, Video* (London: Routledge and Kegan Paul, 1982) is an exemplary book in this respect. Ellis is one of the first to describe in detail both the similarities and differences between the cinema and television institutions.

In *The "Nationwide" Audience: Structure and Decoding* (London: British Film Institute, 1980), David Morley begins the work on how a TV text "inscribes" its viewers, positioning its audience through various modes of address. Robert Deming discusses the different theories of "spectator positioning" in television in "The Television Spectator-Subject," *Journal of Film and Video* 37, no. 3 (Summer 1985): 49–63, and John Caughie, "The 'World' of Television," *Edinburg 77 Magazine #2: History/Production/Memory* (London: British Film Institute, 1977), pp. 73–83, deals with subject-positioning in relation to television's regulated "flow," but this essay is more complex and difficult.

Among the works on individual programs (or types of programming) that rely on a psychoanalytic approach, Stephen Heath and Gillian Skirrow, "Television: A World in Action," *Screen* 18, no. 2 (Summer 1977): 7–59, analyzes how a particular type of viewer is constructed by the documentary/interview format. Margaret Morse and Sandy Flitterman discuss relations of sexuality and desire in televised sports (Morse, "Sport on Television: Replay and Display," in *Regarding Television—Critical Approaches: An Anthology,* ed. E. Ann Kaplan, American Film Institute Monograph Series, vol. 2 [Frederick, Md.: University Publications of America, 1983], pp. 44–66), and a particular detective show (Flitterman, "Thighs and Whiskers: The Fascination of *Magnum, P.I.*," *Screen* 26, no. 2 [March/April 1985]: 42–58). Robert Stam's excellent article "Television News and Its Spectator," also in *Regarding Television,* pp. 23–43, uses Metz's "Imaginary Signifier" in a detailed and highly readable discussion of televised news. Both Stam's and Morse's articles are wonderful examples of the psychoanalytic method as applied to TV. Morse also has another article, "Talk, Talk, Talk—the Space of Discourse in Television," *Screen* 26, no. 2 (March/April 1985): 2–15, which discusses TV news, as well as sportscasts and talk shows, in terms of the relations of subjectivity and discourse.

Two articles that apply psychoanalysis to TV, but do so in ways different than those outlined in this chapter, are Beverle Houston, "Viewing Television: The Metapsychology of Endless Consumption," *Quarterly Review of Film Studies* 9, no. 3 (Summer 1984): 183–95; and Marsha Kinder, "Music Video and the Spectator: Television, Ideology and Dream," *Film Quarterly* 38, no. 1 (Fall 1984): 3–15. The former uses Lacanian theory to analyze the way in which television's lack of spectacle works against producing a unifying experience for the viewer, instead instilling in its audience the desire to consume. The latter article discusses the form and institutional setting of music television, focusing on the relations between video and dreaming.

This list of works on television that use psychoanalysis is intended to be suggestive rather than exhaustive; there are other articles that, though not developing a psychoanalytic theory of television viewing, make use of some of the concepts.

General works on psychoanalysis that are relevant to this kind of work are, of course, Sigmund Freud, *The Interpretation of Dreams, Three Essays on the Theory of Sexuality, Jokes and Their Relation to the Unconscious, The Psychopathology of Everyday Life,* and a collection of essays entitled *General Psychological Theory.* Terry Eagleton's chapter on psychoanalysis in *Literary Theory: An Introduction* (Minneapolis: University of Minnesota Press, 1983) provides a useful overview, as does Juliet Mitchell's founding work, *Psychoanalysis and Feminism* (New York: Vintage Books, 1975). Jacques Lacan's writings, *The Four Fundamentals of Psychoanalysis* (London: Hogarth Press, 1977) and *Ecrits: A Selection* (New York: W. W. Norton, 1977) are notoriously difficult, but there are some texts that go a long way toward clarifying and elucidating the issues. Jean Laplanche and Jean-Baptiste Pontalis use an extended dictionary format in their extremely useful book, *The Language of Psycho-Analysis* (New York: W. W. Norton, 1974); Rosalind Coward and John Ellis, *Language and Materialism* (London: Routledge and Kegan Paul, 1977) has a more difficult prose style but provides helpful discussions on "developments in semiology and the theory of the subject"; and Linda Williams's book on Surrealist cinema, *Figures of Desire: A Theory and Analysis of Surrealist Film* (Urbana: University of Illinois Press, 1981) contains excellent summaries of Lacanian concepts, particularly in their relation to film.

FEMINIST CRITICISM AND TELEVISION
E. ANN KAPLAN

The two parts of my title, "feminist criticism" and "television," require brief discussion individually before I link them together. I need to say something about the contexts in which television studies developed in order to account for the historical paucity of feminist approaches, and, because "feminism" does not have a single meaning, I need to discuss the ways in which I define the term.

I will focus mainly on television studies in the United States, but a contrast with the British approach will illuminate, through differences, developments in this country. In the United States, television studies have had even more difficulty than film in being accepted as an academic subject. Film finally obtained such acceptance through its claims to be "art," but no one was willing to make that argument for television. Thus, whereas the study of film was able to find a place in various humanities departments in the 1960s, the study of television was forced to focus mainly on production. It was housed in schools of journalism and communication, which relied heavily on social-science methodologies of the quantifying and positivist type. Several scholars have discussed the limitation of these methodologies.[1] According to David Morley, the methodologies were developed partly in reaction against the Marxist sociological perspective of the German Frankfurt School, which relied on "social theory and qualitative and philosophical analysis."[2]

In Britain television study took a different tack because it was not developed in schools of communication, but rather through organizations like the British Film Institute and places other than the major universities such as art colleges, further education colleges, and polytechnic institutes. British intellectuals outside of the university communities, then, originally developed methods for studying and criticizing television that were not dissimilar from those used in early cultural and film studies.[3]

It is thanks to the journals *Screen* and *Screen Education,* together with the University of Birmingham's Centre for the Study of Contemporary Culture, that work was developed and debated. The specific focus on the teaching of media studies—of television especially—gave rise to a method of study closely related to the experiences of the spectator, the students being the spectators with whom teachers interacted. In other words, these media scholars began to think of the social contexts for their studies because they were examining their teaching methods/strategies, or because, for political reasons, they were interested in issues of class and race. Their approaches were "naturally" concerned with *reception,* because they were addressing students who were often quite young in terms of their reactions to television.

But there were other reasons why media studies in Britain did not develop along American social-science lines. First, media scholars were interested in television as an institution; they were concerned with the relations of power that involved television, with problems of class, and with the psychological impact of mass culture on the general public. Some of their work built on that of the Frankfurt School, although it took rather different directions. Second, in the mid-1970s further work on television as an institution was stimulated by the announcement of the new Channel 4, which was to be devoted to more experimental and risky material. For all the above reasons, British scholars tended to produce empirically based work that was shaped by a leftist ideology; in America, the quantifying method that became dominant was one which assumed an "objective" stance toward what was being analyzed.[4]

However, neither in America nor in Britain was there much feminist work throughout the 1970s (a paper read by Richard Dyer, Terry Lovell, and Jean McCrindle on "Soap Operas and Women" at the 1977 Edinburgh International Television Program was a rare exception). The gap was noticed in 1980 by Susan Honeyford, who suggested that perhaps the dearth of feminist criticism in television in Britain had to do, first, with "the massive dominance of the national broadcast television institutions with their insistence on large audiences," and, second, with "the relatively little academic work or serious critical writing."[5] She noted that two areas of inquiry had recently arisen in connection with the debate about the fourth television channel: the question of women's employment in the television industry and "the question of programmes made by women and often about women or issues which particularly concern women." However, Honeyford's article dealt mainly with the first question and only very generally with the issue of female representation in television programs. Her discussion implied that women working in the field of television documentary were far more constrained than were women independent filmmakers by their production context but had little theoretical understanding of their plight. A later article by Gillian Skirrow discussed efforts to get TV unions to agitate for more jobs for women,[6] but otherwise *Screen* articles on television, until 1981, dealt mostly with broad issues—definitions of TV studies, teaching strategies—or on the contents of specific programs.[7]

How do we explain such a gap, particularly during a time when feminists were developing strategies for analyzing the classical Hollywood cinema? Perhaps one reason is that American scholars who were both female and feminist tended to work in the humanities rather than in the social sciences. If TV studies concentrated on the (traditionally male-dominated) production sphere and involved social-science methods (again largely male-dominated), then the lack of feminist work in those areas is understandable.[8] For instance, between 1940 and 1970 there were very few studies of soap operas—an obvious TV genre for feminist analysis—even from a quantitative perspective; what studies there were did not foreground gender issues.[9] One of the problems with quantitative research is precisely that it pretends to be able to take an "objective" stance, that is, to provide empirical data untainted by any particular set of concerns or interests. By definition, as we'll see, feminism is a "political" position, and feminist research (no matter what type) must look for issues having to do specifically with women and the place they are assigned in society. For this reason, most of the quantitative, content analyses of soaps do not constitute *feminist* research, although their results may be useful for feminist scholars.

In Britain, the feminists interested in television were more activist-oriented and thus focused on changing the institution itself rather than on developing feminist readings of TV texts. Fewer women in Britain than America have academic positions that give them the privileged time to write; much of the film theory developed in Britain, for instance, was the work of women actually engaged in independent filmmaking, so that the theory was closely linked to the practice. But, as I discuss further below, there was no analagous independent television production with the same possibilities for exhibition, because the nature of television as an institution makes such independent production enormously difficult.[10]

In general, a circular effect was set in motion, such that the more feminist theory was developed for film studies, the less room was left for feminist approaches to TV. It is also possible that the low academic standing of television, together with the often poor quality of American programming, made women—who were already having a sufficiently difficult time getting ahead in humanities departments—reluctant to engage with the form. Women were intrested in and constantly watched films, and very often these same women scholars did not watch television. There was little reason for them to turn their attention to television, because there was so much to do in film and the academic gains from television studies were few.

Things did begin to change in the early 1980s, when a significant number of women finally began to study female representation in television. An article by Stephen Heath and Gillian Skirrow, published in *Screen* in 1977, about the British program *World in Action,* was one of the first to apply British theoretical approaches from the 1970s, first worked out for the classical Hollywood film, to television. A model for any future close-reading analysis, the essay focused on "the fact of television itself," and on "the ideological operations developed in that fact," rather than on discussion of any particular (political) positions that the "content" might reveal.[11]

Meanwhile, various graduate programs in film, such as that at UCLA, began to turn their attention to critical analysis of television, and graduate students produced some of the first interesting work.[12] Some people, like Tania Modleski, who were interested in women's popular fiction also began to apply their critical approaches to related kinds of women's popular TV programs such as soaps.

A television conference held in New York in 1980—to which few, if any, women critics were invited—first made me realize the necessity for more work by women on television and female representation and also how little recognized was the work that had been done. The conference, "Perspectives on Television," that I organized at Rutgers in 1981 tried to remedy this gap and managed successfully to bolster interest in approaches to female representation on television informed by psychoanalysis and semiology, some of which will be discussed below.

But first, let me turn briefly to the second term in need of clarification: "feminist." In the areas of literature and film, feminist criticism already had quite a long history by 1980. Feminist literary criticism dates from the late sixties and the pioneering work of Betty Friedan, Germaine Greer, and Kate Millett, all of whom drew upon insights by Simone de Beauvoir, whose 1949 book, *The Second Sex,* was way ahead of its time.[13] The study of images of women in film dates back to work by the National Organization for Women in the late 1960s; to the pioneering journal, *Women and Film,* published from 1970 to 1972; to film journals like *Jump Cut* that made feminist approaches a central part of their format; and, finally, to the emergence in 1976 of a journal, *Camera Obscura,* specifically devoted to feminist film theory. This journal, influenced by French and British film theory, was a major conduit for that work to American students and film researchers.

Along with critical work, there were in the 1970s a whole host of film conferences and screenings through which women could familiarize themselves, not only with recent independent work being done by women, but also with the work of women directors in Hollywood and throughout the world. These events and the development of critical research enabled a subfield of feminist theory to emerge within film studies. Such conferences and screenings could not be set up around television, largely because of its specific institutional mode: its forms of production and exhibition; its situatedness as a predominating commercial mode within the private home; its lack of historical documents by or about women; the absence of any *alternate* mode such as there had always been in film. Partly because of these conditions, a similar set of "feminist" approaches to television were not readily developed.

By the 1980s, then, the word "feminist" had come to mean a variety of things in literary and film research; I will briefly detail these meanings, because understanding the different kinds of feminist work in allied humanities fields will both explain some of the work that scholars have been doing, and suggest future work that we need to begin doing, in television. By the 1980s, the word "feminist" had come to mean different things to different women, depending on what their theory of feminism was. Because feminism is by definition a *political* and *philosophical* term, let me list the most obvious kinds of feminism as they have been developed under these categories. I want to emphasize before I start, however, that I do not intend here to set up a hierarchy in which any one type is meant as "higher" or "superior" to another.

There is a rough chronological sequence to the types listed, but later methods should not be viewed as necessarily replacing earlier ones. Certain developments naturally challenge what went before (they of course only arise because of earlier work), but it seems to me that the challenges can be answered. A second caveat has to do with the inevitably archetypal nature of any attempt at categorizing. Very rarely will we find a piece of feminist criticism that offers a pure illustration of a particular category; I will discuss many theories that combine types or do not fit neatly into any one type. The categories are merely useful as a charting of the terrain—for purposes of clarification and illustration. All of this will be clearer after I delineate the types of feminist criticism.

Thus, thinking *politically,* one might isolate bourgeois feminism (women's concern to obtain equal rights and freedoms within a capitalist system); Marxist feminism (the linking of specific female oppressions to the larger structure of capitalism and to oppressions of other groups—gays, minorities, the working classes, and so on); radical feminism (the designation of women as different from men and the desire to establish separate female communities to forward women's specific needs and desires); and post-structuralist feminism (the idea that we need to analyze the language order through which we learn to be what our culture calls "women," as distinct from a group called "men," as we attempt to bring about change beneficial to women).[14] By briefly reviewing examples of feminist work on television, I will show that scholars develop a critical method according to their political definition of feminism.

But first a word about the *philosophical* definition of feminism. There are two main philosophical positions—"essentialist" and "antiessentialist"—that have been much discussed in recent American scholarship.[15] Although the distinction is problematic, it makes discussion simpler. One could view the first three political definitions of "feminist" as falling under the category of essentialist feminism, whereas the fourth reflects antiessentialist feminism. This is *not* to deny that there are crucial differences among the first three types; I will be pointing these out in the course of my review. But the broader categories set up a more general and abstract distinction among the kinds of television criticism produced by scholars operating from one or the other of the philosophical definitions.

Essentialist feminism assumes a basic "truth" about women that patriarchal society has kept hidden. It assumes that there is a particular group—"women"—that can be separated from another group—"men"—in terms of an essence that precedes or is outside of culture and that ultimately has to have biological origins. The essential aspects of women, repressed in patriarchy, are often assumed to embody a more humane, moral mode of being, which, once brought to light, could help change society in beneficial directions. Female values become a standard for critiquing the harsh, competitive, and individualistic "male' values that govern society; they offer an alternate way not only of *seeing* but of *being* that threatens patriarchy. Feminists who subscribe to this theory believe that female values, because of their essential humaneness, should be resurrected, celebrated, revitalized. Marxist feminists would, in addition, focus on the way social structures and the profit motive have prevented humane female values from becoming dominant, and radical feminists would emphasize that the silencing of the female voice results from male domination, forced heterosexuality, the insistence on the bourgeois, nuclear family, and so forth. Liberal essentialist feminists remain largely reformist rather than revolutionary and are content merely to assert women's full rights to whatever our society has to offer.

Antiessentialist feminists view things differently, although the philosophical approaches are not necessarily as incompatible as they might at first seem. Antiessentialist theorists attempt to understand the processes through which female subjectivity is constituted in patriarchal culture, and they do not find an "essential" femininity behind the socially constructed subject. The "feminine" is not something outside of, or untouched by, patriarchy, but integral to it. Antiessentialist theorists are concerned with the links between a given sex-identity and the patriarchal order, analyzing the processes through which sexuality and subjectivity are constructed at the same time. We are all agreed by now that we can change *sex-roles*—many Western societies are catching up with the Eastern block in enacting such changes. But antiessentialists argue that for such changes to take a firm hold, to have any more than a merely local, fashionable, and temporary change, we have to understand more about how we arrive at sex-identity in the first place. (The fact that sexism remains a problem in Eastern and other Communist countries attests to the fact that social changes are not sufficient in and of themselves.) If the goal is to get beyond the socially constructed definitions of man/woman or masculine/feminine, anties-

sentialists argue, we need to know precisely how those social constructions are inscribed in the processes of becoming "human," and this inevitably entails moving into the psychoanalytic terrain.

It is significant that between 1963 and about 1980 American feminists thought mainly in essentialist terms and in most disciplines (except film studies) produced work within the frame of the first three
political positions. With the recent influence of a particular type of
French feminist theory [feminisms], scholars have begun to use the
psychoanalytic/semiotic/post-structuralist theories underlying continental work. These theories permit us to look at the various kinds of
feminisms a bit differently. Julia Kristeva's account of the three broad
stages of feminism, seen in a loose historical manner, provides a scheme
for distinguishing types of feminist criticism that combines the political and philosophical categories outlined above.[16] Although there is a
broad developmental aspect to the account of critical methods—a particular kind, as I noted, may be seen as arising out of questions not
answered by a previous approach—it is also true that a new approach
does not invalidate or eliminate earlier ones. In fact, in the eighties we
find that most types of feminist criticism are still being produced con-
currently, where they have not been incorporated into methods that
combine earlier approaches with new ones. The process is synthetic or
dialectical, rather than negational.

In what follows, I provide examples of criticism produced within
each of the three categories of feminism, as defined by Kristeva; as far
as possible, I use soap opera studies to illustrate each method and to
highlight similarities and differences.

Kristeva's analysis relies on Jacques Lacan's theories of the way the
subject is constructed in a patriarchal language order (which Lacan
calls "the Symbolic") and in which woman is normally relegated to the
position of absence, or lack. Lacan's distinction between the Imaginary and the Symbolic, discussed in Sandy Flitterman-Lewis's chapter
on psychoanalytic criticism, is central to the different kinds of feminism being discussed here. For Lacan, the Imaginary *proper* lacks
gender specificity—or rather, it brings both genders into the feminine
through the illusory sense of being merged with the mother. What
Lacan calls the "Mirror Phase" (the moment when the child first sets
up a relationship to its image in the mirror) marks an awareness that
the sense of oneness with the mother is illusory. The child begins to be
aware of the mother as an object distinct from itself (the mirror con-

tains an image of the mother holding the child); it also recognizes its "mirror" self (which Lacan calls an Ideal Imago) as an entity distinct from itself. The subject is thus constituted as a *split* subject (that is, both mother and nonmother, this side of the mirror and within the mirror). It is important that the Ideal Ego constructed during this Mirror Phase is not entirely on the side of the Imaginary; the child unconsciously incorporates the image of the mother as *another image;* it begins to symbolize its own look as that of the Other, and to set in motion the desire for the mother (displaced, as we'll see, into a desire for what she desires) that will persist through its life.

This recognition of the mother as the Other is, according to Lacan, a universal experience and one that is essential for the human-to-be to, in fact, become *human.* The mother-child dyad must be interrupted by the language order ("me"/"not-me") if the child's development is to move beyond the level of the Imaginary. The Mirror Phase thus prepares the child for its subsequent entry into the realm of the Symbolic (by which Lacan means the language and other signifying and representational systems, such as images, gestures, and sound), in which the child takes up its position as a "sexed" being (it recognizes various subject positions such as "he," "she," "you," "it"). Because signifying systems are organized around the phallus as the prime signifier, the woman occupies the place of lack or absence. The boy and girl, thus, find themselves in vastly different positions vis à vis the dominant order once they enter the realm of the Symbolic.

The problem for the girl is in being positioned so as to identify with the mother, which means desiring what the mother desires—the phallus. This desire has nothing to do with anything essential or biological but everything to do with the way that the Symbolic is organized. Lacan's system, particularly as used by Kristeva, frees us from the tyranny of the biological. It also enables us to see that some conventions that certain stages of feminism conceived of as due to "nature" are in fact socially constructed.

Kristeva's first stage of feminism is that in which women demand equal access to the (patriarchal) symbolic order, desiring equality rather than subjugation. This has two possible results. The first is what we have called "domestic" feminism (largely characteristic of the nineteenth century), in which women valorized the patriarchally constructed "feminine." However, they were likely to see this "feminine" as "natural," and they celebrated the qualities assigned women as morally higher or better than the male values of competition and aggressive individualism. The second is what, above, I call "liberal" feminism, more characteristic of our own times, in which women strive for equality with men in the public work sphere. That is, women demand equal access to jobs, to institutional power (of whatever kind), equal pay for equal work, equal benefits across the board, and also changes in family routines to accommodate their right to demanding careers.

Liberal feminism results in a type of television criticism heavily dependent on quantitative content analysis. TV programs are analyzed in terms of the kinds and frequency of female roles they contain. Such studies might examine the degree to which dramas reflect recent changes in the status of women, their movement out of the home and into the work sphere, the characteristics working women are shown to have, the quality of family life, and the involvement (or not) of men in domestic chores.

A recent example of this type of feminist criticism is Diana Meehan's *Ladies of the Evening*.[17] Meehan's aim is to provide "specific and accurate descriptions of television characters and behaviors and some index of change over time." She assumes that television's presentation of women characters are "reflections of women's lives, implicit endorsement of beliefs and values about women in a very popular forum" (p. vii). She combines the quantitative approach—"counting the number of female characters or female heroes, the numbers of times that situation comedy jokes were at the expense of a female or that dramatic acts of violence were committed by women or against women"— with a more qualitative approach so that she can address "questions about women characters' power and powerlessness, vulnerability and strength" (p. viii). In addition, Meehan uses what might be called a comparative approach to determine the degree to which female characters are representative of the female population in society. This involves "considering female characters as real people" (p. viii). Later, Meehan explains her assumption that "viewers . . . evaluate the behavior of others as appropriate or inappropriate compared with television models, and life and its television versions become even more interrelated" (p. 4).

The heart of Meehan's book is her ambitious attempt, not only to isolate and study occurrences of a whole series of female roles (the imp, the goodwife, the harpy, the bitch, the victim, the decoy, the siren, the courtesan, the witch, the matriarch), but also to show the changes in each image from 1950 to 1980. Her findings are too numerous to detail here (readers are referred to chapter 13 in her book), but one main conclusion is that "the composite impression of the good-bad images was a forceful endorsement of a secondary position for women, a place in the world as selfless, devoted adjuncts to men" (p. 113). In addition, Meehan notes that "any other female stance was, at best, an irritation, an interruption, and at worst a threat to world order, a destructive force." She concludes that "except in the rarest of cases, expression of female autonomy, even expression of her own sexuality, was potentially harmful or dangerous" (p. 113).

Another of Meehan's findings concerns occupational role models other than the housewife. Women, Meehan shows, were nearly always in service occupations, whereas males were shown in "controlling occupations." In effect, "the television rendition of the working woman's role is a copy of its portrayal of the housewife." Like the housewife, the working woman is dependent on a male for supervision and direction, and further, she mirrors the domestic model in the kinds of work activities she performs. Finally, there is the familiar dichotomy between homemaker and wage earner, "relegating the home to the housewife and leaving the workplace a single's domain" (p. 123).

Meehan ends her study by observing that "American viewers have spent more than three decades watching male heroes and their adventures, muddied visions of boyhood adolescence replete with illusions of women as witches, bitches, mothers and imps. Television has ignored the most important part of women's lives—their concepts, sensations, aspirations, desires, and dreams. It's time to tell the stories of female heroes—heading families, heading corporations, conquering fears, and coping with change. Good models are needed to connect women to each other and to their society" (p. 131).

This kind of feminist criticism documents what we have come to understand as a prevalent way of imaging women in popular culture, but unfortunately it does not tell us much about how these images are produced (a study that might help us understand their continuity— with small changes—over a thirty-year period) or about exactly how these images *mean,* how they "speak" to the female viewer. We are left with a vague notion of "positive," as opposed to "negative," images of women, and of a standard—the autonomous, self-fulfilling, self-assertive, socially and financially "successful" woman—against which the images are judged to be either positive or negative. It was to contradict just such a way of reading images of women in film that work by Laura Mulvey, Claire Johnston, and others, referred to earlier, was undertaken.

Meehan's model has serious problems. First, it represents the human consciousness as a tabula rasa upon which TV images are graven. Images are seen as models that viewers imitate because they "read" them as real people. Second, it is assumed that this process of imitation is analogous to that which takes place in the family where the child models its personality on that of its parents. What this view obviously leaves out is that fictive characters are *not* real people, and therefore viewers are forced to take a very different position toward them. Also, the processes through which children "identify" with significant adults in their lives is enormously complicated and involves (as we saw in discussing Lacan) the unconscious and the language order in which children are placed. The viewer exists in a dynamic relationship to other people and to the screen image, bringing an already complex unconscious to reception, and in a certain sense being "constructed" as a subject in the processes of reception.

Occasionally, Meehan seems to offer a glimpse of this level of interpretation, as when she notes that "television has never been simply a reflection of society, as evident by the variety and abundance of content which grossly distorts the experience of viewers. The distortion can be attributed to the aspect of television content that is fantasy" (p. 113). She goes on to talk briefly about fantasy as an expression of myth, with a brief reference to Carl Jung, and she mentions the fact that the fantasies that TV shows are basically male ones, although she notes, without exploring the idea, that they also appeal to women (p. 125). But this level of exploration is mentioned only at the very end of Meehan's study and is in opposition to her main theory that "the reality of the images is evident in the recognizable similarities between the action and events of characters and the experiences of the viewers; the goodwife, harpy and bitch display social behavior we encounter in society" (p. 113).

It is clear that, within Kristeva's Lacanian paradigm, such studies show women demanding equal access to the (patriarchal) Symbolic. The "stories" Meehan says it is time "to tell" envision a society in which women are incorporated into the masculine public sphere as "heading families, heading corporations." Women are to be seen as no different than men, but this really means that women are "to become men." The position fails to take into account that such a move demands woman's complete surrender to the patriarchy and its values, norms, and ways of being. As I will explain later on, it implies that woman must replace being defined by the phallus with her identification *with* the phallus; although this may be important as a transitional phase, it should not be seen as an end in itself.

Before moving on to Kristeva's second stage of feminism, it is necessary to insert a category—pre-Althusserian Marxist feminism—not explicitly mentioned by her, perhaps because she did not consider it sufficiently different from the stage just discussed or because it was not specific enough to feminism per se. However, in the early 1970s, this was an approach developed by some American scholars in media and literary studies.

Marxist feminist critics of this early kind construct television as a critical object in relation to the larger economic and industrial power networks in which it is embedded. Relying on the long tradition of what we might call stage-one Marxist criticism (see Mimi White's chapter and bibliography in this volume), feminists who are this kind of Marxist look at how television's status as an explicitly capitalist institution affects what images of women are portrayed. Such Marxist-feminist researchers stress the production of the woman-viewer as a consumer, which emerges from television's need—as a commercial, profit-making institution—to sell objects along with entertainment. But television's reliance on constructing numbers of viewers as commodities involves reproducing female images that accommodate prevailing (and dominant) conceptions of "woman," particularly as these satisfy certain *economic* needs.

Pre-Althusserian Marxist feminists, then, are interested in how women as a group are manipulated by larger economic and political concerns outside of their control. Thus, narratives might construct images of the working woman if society needs women in the work force; alternatively, they might represent woman as content to be a housewife when that is economically beneficial. The approach involves content analysis not that dissimilar from the previous "liberal" or "reformist" feminism, but the ends are different because the Marxist discussion, unlike the other, always takes place within the context of television as a profit-making, capitalist concern.[18] (More recent work by Marxist feminists using an Althusserian and epistemological framework takes into account the complexities of determinations and argues that representations cannot only be explained by the profit motive. I include such work in the post-structuralist feminist category.)

Lillian Robinson's "What's My Line? Telefiction and Women's Work" is an excellent example of this early Marxist feminist approach to television.[19] Written in 1976, before there was much feminist work on television at all, the article deals with the contrast between the image of working women in television serials (including soaps) and the actual situation of women as workers in society. "TV fiction," Robinson argues, "has developed a set of myths specific to women and work elaborating on the themes of whether and why women enter the job market, what occupations they engage in, how they typically perform there, how they interact with the people with and for whom they work." Turning, like Meehan, to "actual working women" to "test" the images, Robinson also finds distortions; whereas TV images collapse women's work identity and situation into sexuality, the real working woman sees herself as a person, "both worker and woman—with a job, a boss, a paycheck, and a set of working conditions, not a complex of sex roles involved in a workplace" (p. 312).

Unlike Meehan, Robinson accounts for these distortions by defining television as "a branch of something called the entertainment industry," which, Robinson says, "implies something [that] is manufactured here, mass-produced by alienated labor for the consumers who constitute its mass audience" (p. 313). However, Robinson refuses to fall into Meehan's trap of conceiving her audience as completely vulnerable to the images provided. She argues, rather, that women do not necessarily accept what they are shown, and that the images are merely "one of the factors that *influence* the consciousness of women" but do not provide the whole story (p. 313).

Robinson proceeds to contrast the statistics relating to women's work in society to TV images of working women. She finds that, despite the trend away from family-based situations in comedy and drama, "the probability of a TV woman's being employed is about half what it would be for her real-life counterpart." In addition, "Motherhood almost always means leaving the work force, which is not too surprising, but marriage itself tends to have the same result" (p. 315). Robinson gives examples from shows like *Days of Our Lives*, *All in the Family*, and *One Day at at Time* to prove her points, and she demonstrates that, in addition to the distorted proportion of women working on TV, there is a large difference between the kinds of work women are seen to do on TV and the work they do in real life.

Robinson first looks at the low-status jobs that TV women do, pointing out that "TV women, both in offices and outside them, tend to be assigned what I think of as 'cutesy jobs,' occupations that require human contact and that place the woman in a series of potentially colorful situations" (p. 324). These jobs, Robinson notes, often entail silly costumes, animals, children, or humiliating situations, and they "create a climate of inference about the general silliness of women's reasons for working, women's jobs, and women's characteristic performance at them" (p. 324). Her Marxist feminist point of view leads her to comment that TV shows carefully do not foreground the fact that these jobs "are normally unproductive and often socially useless" and "that most of them pay minimal wages . . . those jobs arranged through a temporary agency [creating] double exploitation" (p. 324).

Robinson concludes by examining the images of professional women, and she finds, interestingly, that these images are definitely there, "larger than life size and in far greater numbers than the real world, in which most of us still work as typists, waitresses, and saleswomen, would admit" (p. 335). Robinson assumes that the drastic fates that befall these TV professional women are meant as some kind of a warning to real women not to aim so high. This warning, she feels, is premature except insofar as it contributes to three interconnected myths: "that women enjoy a higher status than we feminists claim; that this status has been and may be achieved without fundamental social upheaval; and that having a career nonetheless poses a very real threat to female nature, to individual women's stability, and to institutions like the family that are built on these twin foundations" (p. 335).

Robinson's essay has many of the same strengths and limitations as Meehan's, but it differs in not demanding equal access to the (capitalist) patriarchal Symbolic. Rather, Robinson attempts (although she would not use this language) to show how that very Symbolic exploits and manipulates women workers and, further, constructs images that either belittle women's work or warn women of the deleterious effects of aiming too high. Her objective is to expose the workings of the patriarchal Symbolic rather than with arguing for woman's access to it. But like Meehan, Robinson assumes an essentialist notion that women can resist their exploitation—that they are not socially constructed through the processes of their positioning.

Kristeva's second type of feminism is that in which women reject the male symbolic order in the name of difference, resulting in radical feminism. Femininity is not only celebrated by radical feminism but also seen as better and essentially *different*. The focus is on women-identified women, on striving for autonomy and wholeness through communities of women, or at least through intense relating to other women. Radical feminist criticism might be concerned with TV's depiction of traditional family life as the solution for all ills, with the forced heterosexual coupling in most popular narratives, or with the discrepancies between images of marriage in popular culture and in real life. The failure of popular culture to address women's positive ways of relating to one another and the portrayal of men as "naturally" dominant might also be issues.

Carol Aschur's pioneering essay "Daytime Television: You'll Never Want to Leave Home," written in 1976 (under the name "Lopate"), shows traces of this radical position. Aschur's opening discussion of game shows exposes the infantilized positioning of women vis à vis the "inevitable male M.C."[20] "The M.C.," Aschur notes, "is the sexy, rich uncle, the women, preadolescent Lolitas," and she comments on the way the MC exploits his position to fondle the women and to receive their embraces for the free gifts they get in this unreal, bountiful world (p. 72). The game shows, Aschur reveals, "recreate and transform women's general economic powerlessness as well as their role as consumers." They are shown as dependent for money on men, who control the spending power although women actually make the purchases. Women's power of decision-making is limited to choosing commodities with the money men give them.

Turning to soap operas, Aschur illustrates the two important myths that they propagate; the idea of America "as a country where almost everyone is middle class," and the idea that "the family can be and is, the sole repository of love, understanding, compassion, respect, and sexuality" (p. 74). Soap opera families "portray the idealized lives of families economically headed by professional men," while most women are housewives. Even when they work, they are rarely seen on the job. What Aschur most objects to is, first, the way that the family is set up as central—people, she says, are never allowed to leave the family or to "be alone long enough to develop a real self and thus have a personality that can be known" (p. 79); and second, that soap operas do not reveal "the nonbenign aspect of the power that men hold over women" (p. 81). Soaps misrepresent "real life" in portraying men, like the MC in game shows, as "having the capacity to assist, protect, and give, without retaining the power to dominate that most men potentially have over most women. No soap opera father is a disciplinarian; no husband a wife beater" (p. 81). Aschur concludes that there is more equality between women and men in soaps than in real life or any other dramatic form; and that soaps ultimately function to promise the housewife, confined to her home, that "the life she is in can fulfill her needs." What soaps repress are both her actual loneliness and isolation, and also the idea that it might be precisely through her solitude "that she has the possibility for gaining a self" (p. 81).

In accord with type two feminism, Aschur suggests a need for women to reject the male symbolic order, although, again, she does not use this terminology. That order, as revealed in popular TV shows addressing women, exploits and infantilizes women on the one hand and idealizes the (in fact oppressive) patriarchal family on the other. The implication is that women can and should reject such debasing images, and indeed, such degrading life scenes, in order to find themselves. Autonomy, independence from men, and bonding with other women are suggested, parenthetically, as both possible and desirable. We see here Aschur's essentialism, but it is important to note that she arrives at very different conclusions than does Meehan and differs also from Robinson in suggesting individual rather than social alternatives.

Kristeva's last type, that in which women reject the dichotomy between masculine and feminine as metaphysical and aim at transcendence of the categories of sexual difference—or at least recognition of their cultural construction—is only possible in the wake of the great twentieth-century modernist movements and the postmodernist theories that followed upon those movements. In this stage, scholars analyze the symbolic systems—including the filmic and televisual apparatuses—through which we communicate and organize our lives so as to understand how it is that we learn to be what our culture calls "women" as against what are called "men."

This post-structuralist femininism is often antiessentialist in contrast to the essentialism of the previous three types discussed, although, as we'll see, some of the work combines essentialist and antiessentialist assumptions. Critics with this philosophical orientation have found the work of Laura Mulvey and Claire Johnston central to the formulation of the feminist antiessentialist theory in film studies.[21] These authors were themselves influenced by European theorists discussed in previous chapters—the Russian Formalists; Benjamin and Brecht (Germany); and Barthes, Althusser, Foucault, Lacan, Kristeva, and

Derrida (France). Most important for our discussion here is Mulvey's crucial—and by now much discussed—essay "Visual Pleasure and Narrative Cinema," written in 1975.[22] Mulvey's interest in the commercial Hollywood film as an embodiment of the patriarchal unconscious provoked new interest by women in dominant popular forms. Her analysis involves Hollywood's apparent inscription of the "male," as opposed to any possible "female," unconscious. Drawing on Freud's twin mechanisms of voyeurism and fetishism, Mulvey shows that the dominant Hollywood cinema is built on a series of three basic "looks," all of which satisfy desire in the male unconscious. There is, first, the look of the camera in the filming situation (called the pro-filmic event); although technically neutral, this look is inherently voyeuristic and usually "male," in the sense that a man is generally doing the filming. Second, there is the look of the male figures within the film narrative, and these are organized through shot-countershot so as to make the woman the object of their gaze. Finally, there is the look of the spectator, which imitates (or is necessarily in the same position as) the first two looks. That is, the spectator is forced to identify with the look of the camera, to see as it sees.

Voyeurism and fetishism are mechanisms that the Hollywood cinema uses to construct the (presumedly male) spectator in accordance with the needs of *his* unconscious. Voyeurism is linked to the scopophilic instinct (that is, the male pleasure in his own organ transferred to pleasure in watching other people have sex). Mulvey argues that cinema relies on this instinct, making the spectator essentially a voyeur. Fetishism also comes into play in the cinema, where the whole female body may be "fetishized" in order to counteract the male fear of sexual difference, that is, of castration. Mulvey originally argued that if the spectator is a woman, she has to assume the male position and participate in both mechanisms.

Most critics are agreed that the Hollywood cinema lays out, for our contemplation, unconscious processes that are inaccessible except through psychoanalytic analysis. These theories lead to a set of concerns on the part of feminist scholars employing them that are very different from those addressed by previous essentialist feminists. Following Mulvey, feminist film critics became interested in what she had theorized as an exclusively "male" gaze and in discussing what possible "female" gaze there might be. Scholars soon realized that the theory applied mainly to the central "male" genres—the Western, gangster, adventure, and war films. Women scholars turned to the one film genre that specifically addresses the female spectator—the melodrama—and issues relating to this genre and women viewers are still being actively debated.[23] These scholars began to think about the text-spectator relationship, about exactly how the actual (historical) female viewer (or subject) sitting in the cinema is related to the screen images passing in front of her. Some of this work took a direction similar to that discussed in Robert C. Allen's chapter in this volume on reader-response criticism, and contains some sociological aspects. That is, some work still assumes an interaction between two given entities— the text on the one hand, the reader on the other—whereas other more psychoanalytically oriented approaches assume that the reading-subject is created (or constructed) in the very act of reading—that there is no reader outside of the text, and no text, for that matter, outside of the reader.

Some of the most interesting new feminist work on television, written by people with humanities backgrounds, uses methods developed for studies of the Hollywood film. What this means is that instead of trying to study sex-roles within certain television programs, these studies examine the entire way that television functions, what kind of *apparatus* it is. This apparatus involves the complex of elements including the machine itself (its technological features—the way it produces and presents images); its various "texts"—ads, commentaries, and displays; the central relationship of programming to the sponsors, whose own texts—the ads—are arguably the *real* TV texts (see the discussion of Sandy Flitterman-Lewis's ideas below); and the now various sites of reception, from the living room to the bathroom. Scholars might focus on problems of enunciation, that is, of who *speaks* a text, and to whom it is addressed; or they might look at the manner in which we watch TV, its presence in the home, the so-called "flow" of the programs, the fragmentation of the viewing experience even within any one given program, the unusual phenomenon of endlessly serialized programs; or they might study the ideology embedded in the forms of production and reception, which are not "neutral" or "accidental," but rather a crucial result of television's overarching commercial framework.

As Sandy Flitterman-Lewis discusses in her essay in this volume, one of the still-unresolved issues that nearly every article has to address (whether explicitly or implicitly) is that of the degree to which film theories can apply to the very different "televisual" apparatus. Because feminist film theory evolved in relation to the classical Hollywood cinema, it is particularly important for women who study television to consider how far that theory is relevant to this different apparatus; for example, how well do theories about the "male gaze" apply to watching television, when usually there is no darkened room, where there is a small screen, and where viewing is often interrupted by commercials, by people moving about, or by the viewer switching channels? To what extent is the television spectator addressed in the same manner as the film spectator? Do the same kinds of psychoanalytic processes of subject construction apply? Will semiotics help to illuminate the processes at work? Is there a different form of interaction between the television text and the female viewer than that between the cinema screen and its spectator? What might that relation be?

Let's consider an essay on the soap opera that addresses this last question. Written in 1981 by Tania Modleski, this essay set up the terms of the debate and established a set of interests for much of the work on soaps that followed, even that of scholars who took rather different approaches. Modleski's essay was the first to develop Aschur's suggestion of a relationship between the structure/rhythm/mode of the soap opera and women's work. Modleski was enabled by recent theoretical developments to take the argument further into the realm of the particular psychic demands on woman in the family. Using psychoanalytic arguments from both Nancy Chodorow and Luce Irigaray, Modleski theorizes that "soap operas tend . . . to break down the distance required for the proper working of identification. . . . They point to a different *kind* of relationship between spectator and characters that can be described in the words of Irigaray as "nearness."[24] Modleski uses Chodorow and Irigaray's theories about the mother-daughter relationship to describe the way that the female spectator is socialized to relate to fictional texts: just as a relationship of "nearness" is inevitable in the mother-daughter bonding, which involves a kind of symbiosis, a difficulty of knowing where mother begins and daughter ends, so the female spectator will tend to over-identify with fictional characters and not observe the boundaries that in fact separate her from the image.

Soaps, Modleski goes on to argue, at once rely upon woman's socialized skills in attending to the needs and desires of others and further develop those skills. They also have an episodic, multiple narrative structure that accommodates woman's need to be "interruptable" (she must answer the phone, speak to the neighbor, take in the delivery, attend to the baby, see to the cleaning, ironing, food preparation, and so on) while providing pleasure within the act of teaching "the art of being off center" (p. 71).

Finally, in discussing the alternation between soap narratives and those of commercials (only hinted at by Aschur), Modleski suggests that the two modes address women's dual function as both "moral and spiritual guides and household drudges" (p. 72). Thus, soaps both accommodate the nature of woman's work in the home and make distraction, or interruption, pleasurable. For woman's entertainment, unlike man's, must be consumed on the job, because her "job" is never-ending. Modleski, then, claims that "Woman's popular culture speaks to woman's pleasure at the same time that it puts it in the service of patriarchy, keeps it working for the good of the family" (p. 69).

Just as Modleski makes use of the new interest in psychoanalysis and the screen-spectator relationship to build on work done before, so other scholars built on her essay. Sandy Flitterman-Lewis, for instance, uses the semiotics developed by Christian Metz for film analysis to discuss, in more detail than Modleski or Aschur, the precise nature of the relationship between commercials and the soap drama. She focuses on the processes of enunciation, asking: who speaks the text? To whom is it addressed? Her examination of commercials as texts—that is, as modes of meaning production—reveals that each corresponds to one syntagm (the basic unit of narrative construction that Metz postulated). Soaps themselves, she notes, like any narrative, consist in many syntagms, but *formally* the two kinds of text are similar. Her point is that "far from disrupting the narrative flow of daytime soap opera, commercials can be seen to *continue* it."[25] Commercials, that is, prolong and maintain the overall impulse for narrative that soaps satisfy, while providing units of satisfying closure in an overall form that itself frustrates closure.

In terms of social meanings, Flitterman-Lewis reveals the idealized family present in the commercials as opposed to the families overwhelmed with apparently unresolvable problems in the soaps. The commercials thus function interactively with soaps, setting up a "dialectical alternation between the vision in the soaps and that in the ads" (p. 94). It is this interaction between social meanings in the two sets of narratives that results in commercials having "an important function in shaping society's values" (p. 95).

Flitterman does not explore exactly what these values are that soaps help to shape, but Charlotte Brunsdon, in an analysis of a British soap opera, *Crossroads*, tries to identify them. What she discovers is that instead of being "in the business" of "creating narrative excitement, suspense, delay and resolution," as is the classical Hollywood film, *Crossroads* is, rather, concerned with the ideology of "personal life." In other words, the coherence of the soap does not come from "the subordination of space and time to linear narrativity," but rather from "the continuities of moral and ideological frameworks which inform the dialogue." The serial, that is, takes place within a very circumscribed set of values that provide the norms for everyone's life; even as people are in the process of violating those norms, they are constrained by them and have to learn ultimately to adjust to them or suffer the consequences. According to Brunsdon, the program is "in the business" of "constructing moral consensus about the conduct of personal life. There is an endless, unsettling, discussion and resettling of acceptable modes of behaviour within the sphere of personal relationships."[26]

In addition, Brunsdon is interested in the tension between the subject positions that a text constructs "and the social subject who may or may not take up these positions" (p. 76). Following Paul Willemen, David Morley, Steve Neale (and there are others, like Tony Bennett), Brunsdon stresses that the historical spectator is constructed by a whole range of other discourses, such as motherhood, romance, and sexuality, which will determine her reactions to a text. Looking at program publicity, scheduling, and ads, Brunsdon shows that a female audience for *Crossroads* is implied. She concludes that the address of the soaps is a gendered one that relies on "the traditionally feminine competencies associated with the responsibility for 'managing' the sphere of personal life" (p. 81). Brunsdon is careful to avoid the essentialist trap of claiming such competencies to be "natural" to women; rather, she sees women as socially constructed to possess such skills through inscription in "the ideological and moral frameworks [the rules] of romance, marriage and family life" (p. 81).

Brunsdon's essay is important because it focuses explicitly on the narrative differences between soaps and the classical Hollywood cinema, and on the ideological implications of these differences. The structure of the soap as endless dialogue about personal lives inscribes the viewer in a particular ideological framework regarding the family. This is a positioning quite different from that in the Hollywood film.

It seems to me that exploring these differences in relation to all kinds of TV programs is an important future task. As feminists, we need to explore the degree to which theories worked out for the dominant Hollywood narratives apply to what, above, I called the "televisual apparatus," because the representation of women is produced by the apparatus as much as by the narrative. Indeed, much recent film theory has argued that one cannot make any distinction between the apparatus and the narrative, because it is the apparatus itself that produces certain inevitable "narrative" effects (such as, in film, the forced identification with the look of the camera). This argument, a very complex one, indeed, goes beyond the confines of this chapter.[27] I introduce it simply to highlight what I think is a crucial area for future feminist television research. We need to know how the televisual apparatus is used in any one TV genre to represent the female body—to see what possibilities there are for different kinds of female representation, and how bound by the limits of the apparatus are images of woman on TV. For now, I will refer to my recent work on issues of enunciation, gender address, narrative, and the gaze in rock videos shown on Music Television (MTV) as an example of feminist work that tries to combine analysis of female images in individual texts with attention to their context of production/exhibition and to the televisual apparatus.

Let me begin by discussing the implications of the televisual apparatus for the representation of women on Music Television. MTV is an advertiser-supported, twenty-four-hour cable station for which subscribers do not pay extra. Now owned by ViaCom, Music Television was the brainchild of Robert Pittman, then of Warner Amex Satellite Entertainment Company, who in 1981 had the inspiration of a channel devoted mainly to showing promotional videos provided free by the record companies, on the model of free records provided to radio stations. Confined to a short, four-minute format inserted within the twenty-four-hour flow, rock videos are a unique artistic mode (their song-image form has links with both opera and the Hollywood musical, but as we'll see, it differs in central ways from both). I am interested in the spectator-screen relationship as it is produced by both the visual strategies of individual videos and by the placing of four-minute texts with a series of other four-minute texts, and within a flow that includes other *kinds* of texts, such as product ads, ads for MTV itself and its contests, interviews, music news, and the veejay's comments. How does this "flow" affect the spectator? Is there any particular gender address in this flow? How does this kind of flow particularly affect the *female* spectator?

Let me first say something about the construction of what I have elsewhere called the "decentered" spectator—by this I mean that the very rapid flow of comparatively short segments (parts of a drama, ads, more drama, announcements, ads, news, etc.) produces an experience of fragmentation in the spectator, who is not called upon to concentrate on any particular material for very long.[28] The spectator is constantly diverted to something else instead of being absorbed for a long time, as one might be in a film (shown in a cinema) or by a novel. MTV carries to an extreme this process of fragmentation that is evident in most TV programs. It is particularly true of those that also air on twenty-four-hour stations (like the continuous weather and news channels), but it applies also to those that are "serials" of some kind—that is, continuous segments to be viewed daily (soaps, regularly slotted news that is like a "drama" segment, game shows, etc.).[29]

All of these programs are deployed along a neverending horizontal chain (one moves the dial laterally to receive another program in the box); programs are not discrete units, like Hollywood or other films, to be consumed within a fixed two-hour (or so) limit, presented on an unmoveable screen, and out of the spectator's control; nor are the programs comparable to the novel, which is also clearly bounded or limited by a different sort of "frame" (although more within the reader's control than is the film). TV, in a certain sense, does not have a "frame," because the texts are not bounded in that sort of manner. The texts, rather, resemble an endless filmstrip turned on its side, in which the frames are replaced by episodes. Or, as Peggy Phelan has argued, perhaps a better model is that of Foucault's Panopticon, in which the guard surveys a series of prisoners through their windows.[30] Phelan is interested in setting up the TV producer as the "guard" and the individual TV viewer as the "prisoner" who watches in "a sequestered and observed solitude." But I think the guard metaphor works well also for the spectator's relationship to the various episodes that represent, in Foucault's words, "a multiplicity that can be numbered and supervised"; in fact, for the TV viewer, that desire for plenitude, for complete knowledge, is of course forever delayed, forever deferred. The TV is seductive precisely because it speaks to a desire that is insatiable—it promises complete knowledge in some far-distant and never-to-be-experienced future; its strategy is to keep us endlessly consuming in the hopes of fulfilling our desire; it hypnotizes us by addressing this desire; it keeps us returning for more.

This strategy is particularly evident in Music Television, because the texts are only four minutes long and so keep us forever watching, forever hoping to fulfill our desire in the next one that comes along. The mechanism of "Coming Up Next . . ." that all programs employ, and that is the staple of the serial, is an inherent aspect of the minute-by-minute watching of Music Television. We are trapped in the constant hope that the next video will somehow ultimately satisfy us, and so we go on and on watching and hoping, lured by the seductive and constant promise of immediate plenitude. But all we are actually doing is consuming endlessly.

The question is, to what degree does this decentering televisual apparatus specifically position women? Are women necessarily addressed differently by the apparatus, as was argued (at least in the beginning) for the classical Hollywood film? Is there something inherent in the televisual apparatus that addresses women's social positioning as absence or lack, as was also the case for the Hollywood film?

This question takes me beyond the confines of my topic, but it is possible that what is true for Music Television is true for other TV programs, namely that instead of a more or less monolithic gaze (and a largely male gaze at that) as was found in the Hollywood film, there is a wide range of gazes with different gender implications. In other words, the apparatus itself, in its modes of functioning, is not gender specific per se; but across its "segments," be they soap opera segments, crime series segments, news segments, or morning show segments, we can find a variety of "gazes" that indicate an address to a certain kind of male or female "Imaginary." It is possible that there is frequently a kind of genderless address, and also that people of both genders are able to undertake multiple identifications, depending of course on the program involved.

What this lack of gender specificity implies is that the televisual Imaginary is more complex than the cinematic one and does not involve the same regression to the Lacanian Mirror Phase that theorists discovered in the filmic apparatus, as discussed above. In the case of MTV, for example, instead of the channel evoking aspects of the Lacanian Mirror Phase Ideal Imago—a process that depends on sustained identification with a central figure in a prolonged narrative—it instead evokes issues of split subjectivity, with the alienation that the mirror-

CHART 7·1

Polarized Filmic Categories in Recent Film
Theory that MTV violates:

The Classical Text (Hollywood)	The Avant-garde Text
Realism/Narrative	Non-realist anti-Narrative
History	Discourse
Complicit Ideology	Rupture of Dominant Ideology

image involves (see chart 7-1). In other words, whereas filmic processes seek (especially for the male viewer) to heal, for the duration of
the movie, the painful split subjectivity instituted during the Mirror
Phase, MTV produces the decenteredness that is our actual condition
and that is especially obvious to the young adolescent.

MTV thus addresses the desires, fantasies, and anxieties of young
people growing up in a world in which all traditional categories and
institutions are being questioned. I have elsewhere argued that there
are five main types of video on MTV and that these involve a whole
series of gazes, rather than the broadly monolithic Hollywood gaze
(see chart 7-2 for summary of the types and how the gaze affects female images). The plethora of gender positions on the channel arguably reflects the heterogeneity of current sex-roles, and the androgy-
nous surface of many star-images indicates the blurring of clear lines
between genders characteristic of many rock videos.[31]

Because of both the peculiarities of the televisual apparatus and the
new phase of youth culture produced by the 1960s, most of the feminist methodologies that have emerged in television research so far are
inappropriate for the rock videos on Music Television. This is mainly
due to the sophisticated, self-conscious, and skewed stance that the
texts take toward their own subject matter. It is often difficult to know
precisely what a rock video actually means, because its signifiers are

CHART 7-2

Modes
(All use avant-garde strategies, especially self-reflexivity, play with the image, etc.)

	ROMANTIC	SOCIALLY CONSCIOUS	NIHILIST	CLASSICAL	POSTMODERNIST
STYLE	Narrative	Elements varied	Performance Anti-narr.	Narrative	Pastiche No linear images
LOVE/SEX	Loss and Reunion (Pre-Oedipal)	Struggle for Autonomy Love as Problematic	Sadism/Masochism Homoeroticism Androgyny (Phallic)	The Male Gaze (Voyeuristic, Fetishistic)	Play with Oedipal positions
AUTHORITY	Parent figures (positive)	Parent and public figures: Cultural critique	Nihilism Anarchy Violence	Male as Subject Female as Object	Neither for nor against authority (ambiguity)

PREDOMINANT MTV THEMES

not linked along a coherent, logical chain that produces one unambiguous message. The mode, to use Fredric Jameson's contrast, is that of *pastiche* rather than parody.[32] By this expression, Jameson means that whereas modernist texts often took a particular critical position vis à vis earlier textual models, ridiculing specific stances or attitudes in them or offering a sympathetic, comic perspective on them, postmodernist works tend to use *pastiche,* a mode that lacks any clear positioning toward what it shows or toward any earlier texts that are used.

Jameson's analysis of pastiche has implications for gender representations in videos; it is often unclear who is speaking the rock video text—and therefore whether the male or the female discourse dominates—and the video's attitude toward sex and gender is often ambiguous. One is uncertain, for instance, whether or not a video like John Parr's "Naughty Naughty," or John Cougar Mellencamp's "Hurts So Good," are virulently sexist or merely pastiching an earlier Hollywood sexism. Even in the category that I call "classical" (see chart 7-2), where the gaze is clearly voyeuristic and male, there is a studied self-consciousness that makes the result quite different from that in the dominant commercial cinema.

The ambiguity is usually as prevalent in the videos of songs by female stars as in those of the dominant white male stars featured on the channel. In a forthcoming paper, I argue that "socially conscious" videos by female stars range from those that make a statement one could call "feminist" (Pat Benatar's "Love Is a Battlefield" or Donna Summer's "She Works Hard for the Money") and that have fairly conventional narratives (although they still do not work as logically as the Hollywood narrative), through those that comment on the objectifying male gaze (Tina Turner's "Private Dancer"), to those that attempt to set up a different gaze altogether, or to address some (possible) female gaze (as happens in the Annie Lennox/Aretha Franklin video "Sisters Are Doin' It for Themselves").[33] It is significant that all these types of videos occur infrequently within the twenty-four hour continuous MTV flow, and it is the situation of individual texts within that flow that has implications for gender issues. According to a recent quantitative study of MTV, videos featuring white males monopolize 75 percent of the twenty-four flow. Only 20 percent of MTV videos have central figures who are female (incidentally, the figure is even lower for blacks), and women are typically, like blacks, rarely important enough to be part of the foreground. Jane Brown and Kenneth Campbell assert that "white women are often shown in passive and solitary activity or are shown trying to gain the attention of a man who ignores them."[34] Among those 20 percent of videos with female leads, the number of videos by women that fall into my three categories above is miniscule; those videos rarely, or only briefly, fall into the frequent cycling pattern. The female videos that *are* frequently cycled fall into the first type mentioned—those where the position is ambiguous, where what we might call a post-feminist stance is evident. What precisely I mean by this will be clear from the following discussion; let me merely note that it has to do with the inevitable changes in earlier feminist interventions (of whatever kind) produced by a postmodern culture—that is, one that has already absorbed and incorporated elements of the various [feminisms]. (See note 14 for more discussion of this.)

Let me demonstrate my arguments by looking at a video figuring Madonna, one of the most successful women stars to date. "Material Girl" is particularly useful for discussion because it exemplifies a common rock-video phenomenon, namely the establishment of a unique kind of intertextual relationship with a specific Hollywood movie. Because of this connection, as well as the difficulty of ensuring the text's stance toward what it shows and the blurring of many conventional boundaries, I would put the video in the "postmodern" category in my chart, despite its containing more narrative than is usual for the type.

As is well known, "Material Girl" takes off from the famous Marilyn Monroe dance/song sequence "Diamonds Are a Girl's Best Friend," in *Gentlemen Prefer Blondes*. The sequence occurs towards the end of the film, after Esmond's father has financially severed Monroe from Esmond, forcing her to earn a living by performing. Having finally found her, Esmond is sitting in the audience watching the show. We thus have the familiar Hollywood situation in which the woman's performance permits her double articulation as spectacle for the male gaze (that is, she is the object of desire for both the male spectator in the diegetic audience and for the spectator watching the film). The strategy formalizes the Mirror-Phase situation by framing the female body both within the stage proscenium arch and the cinema screen.

During the sequence in question, which starts with Esmond's astonished gaze at Monroe from the theatre seat (presumably he is surprised anew by Monroe's sexiness), Monroe directs her gaze toward the camera that is situated in Esmond's place. The space relations are thus quite simple, there being merely the two spaces of the stage and of the theater audience. We know that the film is being made under the authorial label "Hawks," that within the diegesis, Monroe and Russell are setting up the action, but that, despite this, the patriarchal world in which they move constrains them and makes only certain avenues available.

When we examine the video inspired by the Monroe dance sequence, we see that the situation is far more complicated. First, it is unclear who is speaking this video, even on the remote "authorial label" level, because, as a general rule, credits are never given. Is it perhaps Madonna, as historical star subject? Is it "Madonna I," the movie-star protagonist within the "framing" diegesis? Is it "Madonna II," the figure within the musical dance diegesis? Is it the poor director who has fallen in love with her image and desires to possess her? If we focus first on the visual strategies and then on the soundtrack, we get different and still-confusing answers to the question.

Visually, the poor director's gaze seems to structure some of the shots, but this gaze is not consistent, as it is in the Monroe sequence. And shots possibly structured by him (or in which he is later discovered to have been present) only occur at irregular intervals. The video begins by foregrounding (perhaps pastiching?) the classical Hollywood male gaze: there is a close-up of the poor director, played by Keith Carradine (the video thus bows again to the classical film), who we soon realize is watching rushes of a film starring "Madonna I." With an obsessed, glazed look on his face, he says, "I want her, George." George promises to deliver as we cut to a two-shot of the men, behind whom we see the cinema screen and Madonna I's image, but as yet we hear no sound from the performance. The camera closes in on her face and on her seductive look, first out to the camera, then sideways to the men around her. As the camera now moves into the screen, blurring the boundaries between screening room, screen, and the film set (the space of the performance that involves the story of the material girl, Madonna II), the "rehearsal" (if that is what it was) ends, and a rich lover comes onto the set with a large present for Madonna I.

This, then, is a desire for the woman given birth through the cinematic apparatus in classic manner; yet although the sequence seems to foreground those mechanisms, it does not appear to critique or in any way comment upon them. In Jameson's terms, this lack of criticism makes the process pastiche rather than parody and puts it in the postmodernist mode. The blurring of the diegetic spaces further suggests postmodernism, as does the following confusion of enunciative stances in the visual track alone. For although the poor director's gaze clearly constructed the first shot-series, it is not clear that his gaze structures the shot where Madonna I receives the present. We still hear the whirring sound of a projector, as if this were still the screening room space; and yet we are *inside* that screen—we no longer see the space around the frame, and thus the viewer is disoriented.

We cut to a close-up of a white phone ringing and a hand picking it up, and we are again confused spatially. Where are we? Whose look is this? There has been no narrative preparation for the new space or for the spectator address: the phone monologue by Madonna I (the only time in the entire video that she speaks) establishes the space as her dressing room. As she speaks, the camera behaves oddly (at least by standard Hollywood conventions), dollying back slowly to the door of her room, to reveal the poor director standing there. Was it, then, his gaze that structured the shot? At the moment the camera reaches him, the gaze certainly *becomes* his, and Madonna I is seen to be its object. The phone monologue that he and the viewer overhears establishes that Madonna I has just received a diamond necklace, and the news causes the poor director to throw his present into the wastebasket that the janitor happens to be carrying out at that moment. It also establishes that Madonna I is *not* the "material girl" of her stage role, because she offers the necklace to her (presumed) girlfriend on the phone.

We now cut back to the stage space that we presume is the film set. This space is ambiguous because the diegesis does not foreground the filming processes and yet there is no audience space. Rather, Madonna II sets up a direct rapport with the camera filming the rock video, and therefore with the TV spectator, deliberately playing for him/her rather than for the men in frame. But the spatial disorientation continues: there is a sudden cut to the rear of a flashy red car driving into the studio; followed by shots of Madonna I's elegant body in matching red dress (knees carefully visible); to her rich lover bending over her; and to her face and apparently dismissive reply. Whose gaze is this? Who is enunciating here? As Madonna I leaves the car, we discover the poor director again, but the series of shots could not have been structured by his gaze.

We cut back to the stage/film set for the most extended sequence of the performance in the video. This sequence follows the Monroe "Diamonds" dance closely and stands in the strange intertextual relationship already mentioned—we cannot tell whether or not the Monroe sequence is being commented upon, simply used, or ridiculed by exaggeration (which sometimes seems to be happening). The situation is further complicated by the fact that *Gentlemen Prefer Blondes* is itself a comedy, mocking and exaggerating certain patriarchal gender roles. Even more confusing, occasional play with the image in the video destroys even the illusion of the stability of the stage/set space; at least once, a two-shot of Madonna II and one of the lovers is simply flipped over, in standard rock video style but in total violation of classical codes that seek to secure illusionism.

Because there is no diegetic audience, the spectator is now in direct rapport with Madonna I's body, as she performs for the TV spectator. There is again no diegetic source of enunciation; the spectator either remains disoriented or secures a position through the body of the historical star—Madonna—who is implied as "producing" the video or simply fixed on as a centering force. This is an issue to be taken up shortly.

Most of the camera work in the dance sequence involves sharp images and either long shots (the camera follows Madonna I's movements around the stage) or straight cuts, but toward the end of the dance rehearsal, the style changes to superimpositions and deliberately blurry shots, suggestive perhaps of a heightened eroticism. The camera allows Madonna I's head to be carried by the men underneath itself, so that only her arm remains in view; after some "dazed" shots, the camera pans left along the edge of the set to discover brown stairs with the poor director standing by them, gazing at the performance. But once again, the sequence was not set up, as it would have been in conventional Hollywood codes, as his structuring gaze; the gaze is only discovered *after* the fact, thus allowing enunciative confusion.

The same disorientation continues in a shot (perhaps a flashforward, although that term suggests precisely the kind of narrative coherence that is missing here) that follows another dance sequence. The poor director is seen bringing simple daisies to a now-smiling and receptive Madonna I, clothed in white (a play on Hollywood signifiers for innocence?), in her dressing room. We cut to the end of the stage performance (there are repeated blurry shots, again signifying—perhaps—sexual delirium), before a final cut to the space outside the studio, where the poor director is seen paying someone for the loan of a car. As Madonna I walks seductively out of the studio, the poor director ushers her into his car. The last shot is taken through the now rain-sodden glass (inexplicable diegetically) and shows their embrace.

This brief analysis of the main shots and use of diegetic spaces demonstrates the ways in which conventions of the classic Hollywood film, which paradoxically provided the inspiration for the video, are routinely violated. The purpose here was to show how, even in a video that at first appears to remain within those conventions—unlike many other videos, whose extraordinary and avant-garde techniques are immediately obvious—regular narrative devices are not adhered to. But the video violates classic traditions even more with its sound-image relations.

This aspect of the video brings up the question of the rock video's uniqueness as an artistic form—namely as a form in which the sound of the song, and the "content" of its lyrics, precedes the creation of images to accompany the music and the words. Although there are analogies to both the opera and to the Hollywood musical, neither form is a direct antecedent for the rock video, in which the song-image relationship is unique. The uniqueness has to do with a certain arbitrariness of the images used along with any particular song; with the lack of spatial limitations; with the (frequently) extremely rapid montage-style editing not found generally (if at all) in the Hollywood musical song/dance sequences; and, finally, with the precise relationship of sound, both musical and vocal, to image. This relationship involves the links between musical rhythms and significations of instrumental sounds and images provided for them; links between the significations of the song's actual *words* and images conjured up to convey that "content"; links between any one musical phrase and the accompanying words, and the relay of images as that phrase is being played and sung.

This is obviously a very complex topic—far beyond my scope here—but let me demonstrate some of the issues in relation to "Material Girl," where things are far simpler than in many videos. We have seen that on the visual track there are two distinct but linked discourses: that involving the poor director's desire for Madonna I (his determined pursuit and eventual "winning" of her); and that of Madonna I's performance of Madonna II, the material girl. These discourses are not hierarchically arranged as in the usual Hollywood film, but, rather, are arranged along a horizontal axis, neither subordinated to the other. In terms of screen time, however, the performance is given precedence.

When we turn to the soundtrack, we find that, after the brief introductory scene in the screening room (a scene, by the way, that is often cut from the video), the soundtrack consists entirely of the lyrics for the song "Material Girl." This song deals with a girl who will only date boys who "give her proper credit," for whom love is reduced to money. Thus, all the visuals pertaining to the poor director-Madonna I love story do not have any correlaton with the soundtrack. We merely have two short verbal sequences (in the screening room and dressing room) to carry the entire second story; in other words, soundtrack and image track are not linked for that story. An obvious example of this discrepancy is the shot of Madonna I (arriving at the studio in the flashy car) rejecting her rich lover: Madonna lip-synchs "That's right," from the "Material Girl" song—a phrase that, in the song, refers to her only loving boys who give her money—in a situation where she actually *refuses* to love the man who is wealthy!

In other words, the entire video is subordinated to the words, with their signifieds that refer in fact only to the stage performance. The common Hollywood musical device of having the dance interlude simply be an episode in the main story seems here to be reversed—the performance is central, while the love story is reduced to the status of a framing narrative. Also significant here is the disjunction between the two stories, the framing story being about a "nice" girl and the performance being about the "bad" girl—but even these terms are blurred by the obvious seductiveness of the "nice" girl, particularly at the end as she walks toward the car in a very "knowing" manner.

We see thus that the usual hierarchical arrangement of discourses in the classical realist text is totally violated in "Material Girl." Although Madonna I is certainly set up as object of the poor director's desire, in classical manner, the text refuses to let her be controlled by that desire. This is achieved by unbalancing the relations between framing-story and performance-story so that Madonna I is overridden by her stage figure, Madonna II—the brash, gutsy material girl. The line between "fiction" and "reality" within the narrative is thus blurred, with severe consequences just because the two women are polar opposites.

In *Gentlemen Prefer Blondes*, on the other hand, no such confusion or discrepancy exists. From the start, Monroe's single-minded aim is to catch a rich man, and she remains fixed on that object throughout. The function of her performance of "Diamonds Are a Girl's Best Friend" is, in part, simply to express what has been obvious to the spectator—if not to Esmond—all along, but also to let Esmond get the idea, were he smart enough. Monroe sings a song that expresses her philosophy of life, but we are clear about the lines between the stage-fiction and the context of its presentation, and Monroe as a character in the narrative. Part of the confusion in the Madonna video comes about precisely because the scene of the performance is not made very clear and because the lines between the different spaces of the text are blurred.

To anticipate John Fiske's discussion in the following chapter, the situation in "Material Girl" is even more problematic because of the way that Madonna, as historical star-subject, breaks through her narrative positions with her strong personality, her love of performing for the camera, her inherent energy and vitality. Madonna searches for the camera's gaze and for the TV spectator's gaze that follows it because she relishes being desired. The "roles" melt away through her unique presence, and the narrative incoherence discussed above seems resolved in our understanding that Madonna herself, as historical subject, is the really "material girl"!

It is perhaps Madonna's success in articulating and parading the desire to be desired in an unabashed, aggressive, gutsy manner (as against the self-abnegating urge to lose oneself in the male that is evident in the classical Hollywood film) that attracts the hordes of twelve-year-old fans that idolize her and crowd her concerts. The amazing Madonna "look alike" contests (as in a recent Macy's campaign in New York) and the successful exploitation of the weird Madonna style of dress by clothing companies attest to this idolatry. It is significant that Madonna's style is a far cry from the conventional "patriarchal feminine" of the women's magazines—it is a cross between a bordello queen and a bag lady. Young teengers can use her as a protest against their mothers and the normal feminine while still remaining very much within those modes (in the sense of spending much money, time, and energy on their "look"; the "look" is still crucial to their identities, still designed to attract attention, even if provocatively).

In some sense, then, Madonna represents the post-feminist heroine in that she combines unabashed seductiveness with a gutsy independence. She is neither particularly male- nor female-identified and seems mainly to be out for herself. This post-feminism is part of a larger postmodernist phenomenon that her video also embodies in its blurring of hitherto sacrosanct boundaries and polarities of the various kinds discussed. The usual bipolar categories—male/female, high art/pop art, film/TV, fiction/reality, private/public, interior/exterior—simply do not apply to "Material Girl."

This analysis of "Material Girl" has shown the ambiguity of enunciative positions within the video that, in turn, is responsible for the ambiguous representation of the female image. The positioning of a video like "Material Girl," moreover, within the twenty-four-hour flow of this commercial channel ensures that it is *this* sort of ambiguous image that appears frequently, as against the other female images mentioned, which are only rarely cycled. That this sort of postmodern image is seen as the most marketable one is evident from the fact that the same post-feminist image dominates the ads interspersed among the videos.

The whole televisual apparatus thus contributes to the prevalence of the ambiguous female image. To summarize: the main force of MTV as a cable channel is consumption on a whole variety of levels, ranging from the literal (selling the sponsors' goods, the rock stars' records, and MTV itself) to the psychological (selling the image, the "look," the style). MTV is, more obviously than other programs, one nearly continuous advertisement, the flow merely being broken down into different *kinds* of ads. More than other programs, then, MTV positions the spectator in the mode of constantly hoping that the next ad-segment (of whatever kind) will satisfy the desire for plenitude—the channel keeps the spectator in the consuming mode more intensely because its items are all so short.

Because the mode of address throughout is that of the ad, the channel relies on engaging the spectator on the level of unsatisfied desire. This remains in the psyche from the moment of entry into the Lacanian Symbolic and is available for channeling in various directions. Given the organization of the Lacanian Symbolic around the phallus as signifier, it is not surprising that MTV basically addresses the desire for the phallus (and for the authority and power it represents) remaining in the psyche of both genders. This partly accounts for the dominance of the channel by videos featuring white male stars.

Nevertheless, as my chart shows, the male gaze is not monolithic on MTV; the televisual apparatus enables the production of a variety of different gazes due to the arrangement of a series of short, constantly changing segments in place of the two-hour continuous film narrative, or the (usually) single book-length or theatrical narrative. There is no possibility within the four-minute segment (others are shorter) for regression to the Freudian Oedipal conflicts in the manner of the classical narrative. What we have instead is a semi-comical play with Oedipal positions in the postmodern video, or a focus on one particular mode in the Oedipal complex in some of the other video types outlined in the chart.

The implications of all this for a feminist perspective need close analysis. Feminism has traditionally relied on a liberal or leftist humanist position, as was clear in my earlier delineation of major types of historical feminism. Baudrillard has suggested that the TV screen symbolizes a new era in which the old forms of production and consumption have given way to what he calls a new "universe of communication." This new universe, unlike the old one (which he argued involved striving, ambition, and struggles of the son against the patriarchal Father) relies on connections, feedback, and interface; its processes are narcissistic and involve constant surface change.[35] If indeed the televisual apparatus manifests a new stage of consciousness in which liberal/left humanism no longer has a place, this affects a majority of feminist positions; and feminism needs to address the changed situation. Gender has been one of the central organizing categories of what Baudrillard calls the old "hot" (as against the new "cold") universe, but this distinction may be lost in the new era, with as-yet unclear (and not necessarily progressive) results.

Feminists need to explore television's part in the changed and still-changing relationship of self to image. This change began at the turn of the century with the development of advertising and of the department-store window; it was then further affected by the invention of the cinematic apparatus, and television has, in turn, produced more changes. The television screen now replaces the cinema screen as the central controlling cultural mode, setting up a new spectator-screen relationship that I have begun to analyze in the work on MTV. For MTV constantly comments upon the self in relation to image (especially the TV image), to the extent that this commentary can be seen as its main "content." The blurring of distinctions between a "self" and an "image"—or the reduction of the old notion of "self" to "image"—is something for feminists to explore, even as we fear the coming of Baudrillard's universe of "simulacra." By this, Baudrillard means a world in which all we have are simulations, there being no "real" external to them, no "original" that is being copied.[36] It would be as if all were reduced merely to exteriors and there were no longer any interiors.

The reduction of the female body to merely an "image" is something that women have lived with for a long time, and the phenomenon has been extensively studied by feminist film critics. But these studies always somehow assumed an entity that could possibly be constructed differently. The new postmodern universe—with its celebration of the look, the surfaces, textures, the self-as-commodity—threatens to reduce everything to the image/representation/simulacrum. Television—with its decentered address, its flattening out of things into an endless, unbounded, unframed network or system, the parts of which all rely on each other—is an apparatus urgently requiring more thorough examination, particularly in relation to its impact on women.

NOTES

1. E. Ann Kaplan, introduction, and William Boddy, "Loving a Nineteen-Inch Motorola: American Writing on Television," in *Regarding Television—Critical Approaches: An Anthology*, ed. E. Ann Kaplan, American Film Institute Monograph Series, vol. 2 (Frederick, Md.: University Publications of America, 1983), pp. xi–xxiii, 1–11; Robert C. Allen, *Speaking of Soap Operas* (Chapel Hill: University of North Carolina Press, 1985), chap. 2, especially pp. 40–44; David Morley, *The "Nationwide" Audience: Structure and Decoding* (London: British Film Institute, 1980), pp. 1–5.

2. Morley, *The "Nationwide" Audience*, p. 1.

3. Recently, a few scholars (such as Charlotte Brunsdon) have undertaken work on television from within the university.

4. For British work of this kind, see Morley, *The "Nationwide" Audience*, and The Glasgow University Media Group, *Bad News* (London: Routledge and Kegan Paul, 1976) and *More Bad News* (London: Routledge and Kegan Paul, 1980). For typical American work in the social-science mode, see G. Comstock et al., *Television and Human Behavior* (New York: Columbia University Press, 1978).

5. Susan Honeyford, "Women and Television," *Screen* 21, no. 2 (1980): 49.

6. Gillian Skirrow, "Representation of Women in the Association of Cinematograph, Television and Allied Technicians," *Screen* 22, no. 3 (1981): 94–102.

7. See, for example, John Caughie, "Progressive Television and Documentary Drama," *Screen* 21, no. 2 (1980): 9–35; Michael Poole, "The Cult of the Generalist: British Television Criticism 1936–83," *Screen* 25, no. 2 (March/April 1984): 41–62; Tony Pearson, "Teaching Television," *Screen* 24, no. 3 (May/June 1983): 35–43.

8. It was only in the wake of the work by feminist theorists in the 1980s that a book did emerge from a "communications" scholar, who also happened to be

male! That scholar was the editor of this text, Robert C. Allen; the book, *Speaking of Soap Operas*. I will discuss this work later on.

9. Bradley S. Greenberg et al., "The Soaps: What's On and Who Cares?" *Journal of Broadcasting* 26, no. 2 (Spring 1982): 519–35.

10. As I discuss later on, women (at least in America) have been making independent videos since the 1970s (a recent series at Artist Space in New York demonstrated the variety and creativity of much of this early work) and continue to make them today. There remains the problem of suitable sites of exhibition; film is still the more familiar form; the prevalence of film courses in academic institutions has further provided a suitable context of exhibition still denied to independent television. Helen Baehr, "A Feminist Reworking of Media Studies," paper presented at the Nordic Conference on Women and the Mass Media, 7–11 April 1986.

11. Stephen Heath and Gillian Skirrow, "Television: A World in Action," *Screen* 18, no. 2 (Summer 1977): 7–59.

12. I am thinking here of work by Cathy Schwichtenberg on *The Rockford Files*, Rebecca Baillin on *Charlie's Angels*, and others. Ellen Seiter, at Northwestern University, was another "pioneer" in the area of soap operas.

13. See Simone de Beauvoir, *The Second Sex*, trans. H. M. Parshley (1949; reprint, Harmondsworth, Eng.: Penguin, 1972); Kate Millett, *Sexual Politics* (New York: Doubleday, 1969); and Germaine Greer, *The Female Eunuch* (London: MacGibbon and Kee, 1970).

14. One could add yet one more political feminist stance—the postmodern— but I have not yet seen this approach applied to television. The concept of the postmodern, as developed particularly by Fredric Jameson and Jean Baudrillard, involves the blurring of hitherto sacrosanct boundaries and polarities, the elimination of any position from which to speak or judge. It involves the obliteration of any distinction between an "inside" and an "outside," the reduction of all to one level, often seen as that of the *simulacra*. There is no longer a realm of the "real" versus that of "imitation" or "mimicry," but rather a level in which there is *only simulation*. Post-structuralist feminism still honors the concept of a feminist position, even while doubting that the (patriarchal) "feminine" can be subverted. It still problematizes the issue of searching for alternate female positions; it still seeks such positions. Postmodernist [feminisms] (the word must be put in brackets because the position is ultimately a *post*-feminist one) rather defines feminism itself as a concept, as an essentializing term, as looking back to the individualist framework of early feminist movements. It therefore refuses to honor "feminist" as a concept but is, rather, interested precisely in elaborating/dwelling upon/working through the problems with the concept. Thus, the postmodernist position is still concerned with feminism, but now from the perspective of exposing its limits, of arguing for the need to think through other categories.

15. See a recent article of mine, "The Hidden Agenda: A Review of *Re-Vision: Essays in Feminist Criticism*," *Camera Obscura*, no. 24/25 (Fall 1985): 235–49. See also my forthcoming essay reviewing feminist film criticism gen-

erally, "Feminist Film Criticism: Current Issues," in *Studies in the Literary Imagination*. In both articles I discuss the essentialist/antiessentialist debate in some detail.

16. Julia Kristeva, "Women's Time," trans. Alice Jardine and Harry Blake, *Signs* 7, no. 1 (Autumn 1981): 13–35.

17. Diana Meehan, *Ladies of the Evening: Women Characters of Prime-Time Television* (Metuchen, N.J.: Scarecrow Press, 1983). Subsequent references are cited in the text.

18. For an excellent example of this kind of work on media images before the invention of television, see Maureen Honeywell, *Creating Rosie the Riveter* (Amherst: University of Massachusetts Press, 1984).

19. Lillian Robinson, "What's My Line? Telefiction and Women's Work," in *Sex, Class and Culture* (Bloomington: Indiana University Press, 1978), pp. 310–44. Subsequent references are cited in the text.

20. Carol Lopate, "Day-Time Television: You'll Never Want to Leave Home," *Feminist Studies* 4, no. 6 (1976): 70–82. Subsequent references are cited in the text.

21. These theories, too complex to review here, have been thoroughly discussed in two recent books on women in film: E. Ann Kaplan, *Women and Film: Both Sides of the Camera* (London: Methuen, 1983); and Annette Kuhn, *Women's Pictures: Feminism and Cinema* (London: Routledge and Kegan Paul, 1982). Please refer to these for more details.

22. Laura Mulvey, "Visual Pleasure and Narrative Cinema," *Screen* 16, no. 3 (Autumn 1975): 6–18.

23. See, for example, the debate over various readings of *Stella Dallas*, in *Cinema Journal* 24, no. 2 (Winter 1985), and 25, no. 1 (Fall 1985): 51–54.

24. Tania Modleski, "The Rhythms of Reception: Daytime Television and Women's Work," in *Regarding Television*, pp. 67–75. Subsequent references are cited in the text.

25. Sandy Flitterman, "The *Real* Soap Operas: TV Commercials," in *Regarding Television*, pp. 84–96. Subsequent references are cited in the text.

26. Charlotte Brunsdon, "*Crossroads*: Notes on Soap Opera," in *Regarding Television*, pp. 76–83. Subsequent references are cited in the text.

27. For further discussion of these complex issues, see the chapter by Sandy Flitterman-Lewis in this volume.

28. For details of these arguments, see E. Ann Kaplan, "A Post-Modern Play of the Signifier? Advertising, Pastiche and Schizophrenia in Music Television," in *Television in Transition*, ed. Phillip Drummond and Richard Paterson (London: British Film Institute, 1985), pp. 146–63; Kaplan, "Sexual Difference, Pleasure and the Construction of the Spectator in Music Television," *Oxford Literary Review* 8, no. 1/2 (1986): 113–23; and Kaplan, "History, Spectatorship, and Gender Address in Music Television," *Journal of Communication Inquiry* 10, no. 1 (Winter 1986): 3–14.

29. See Robert Stam, "Television News and Its Spectator," in *Regarding Television*, pp. 23–44.

30. Peggy Phelan, "Panopticism and the Uncanny: Notes Toward Tele-

vision's Visual Time," unpublished paper to be presented at the annual meeting of the Modern Language Association, December 1986.

31. For more details of these points, see works by E. Ann Kaplan cited in n. 28 above, and also Kaplan's forthcoming book, *Rocking around the Clock: Consumption and Postmodern Culture in Music Television* (London: Methuen, 1987).

32. Fredric Jameson, "Postmodernism and Consumer Culture," in *The Anti-Aesthetic: Essays on Postmodern Culture*, ed. Hal Foster (Port Townsend, Wash.: Washington Bay Press, 1983), p. 113.

33. See E. Ann Kaplan, "Whose Imaginary? The Televisual Apparatus, the Female Body and Textual Strategies in Select Rock Videos on MTV," in *Cinematic Pleasure and the Female Spectator*, ed. Deidre Pribram (London: Verso, forthcoming 1987).

34. Jane D. Brown and Kenneth C. Campbell, "The Same Beat but a Different Drummer: Race and Gender in Music Videos," *Journal of Communication* (Spring 1986): 15.

35. Jean Baudrillard, "The Ecstasy of Communication," in *The Anti-Aesthetic*, p. 127.

36. Jean Baudrillard, *Simulations*, trans. Paul Foss et al. (New York: Semiotext(e), 1983).

FOR FURTHER READING

As noted in the essay, there has been comparatively little work on the representation of women on television altogether, and even less from perspectives other than the quantifying, social-science ones. The main body of feminist work in media studies has been on the Hollywood film, with work on women and the avant-garde, experimental, or documentary film taking second place. As also noted in the essay, methodologies worked out for the Hollywood film do not automatically apply to the different televisual apparatus. Nevertheless, it might be useful for students to familiarize themselves with a modest amount of feminist film theory, because much of the best feminist work on television starts off from those positions and relies on similar theories (i.e., semiotics, psychoanalysis, post-structuralism), revising them according to the specificity of television.

Three recent books provide overviews of feminist film theory: Annette Kuhn, *Women's Pictures: Feminism and Cinema* (London: Routledge and Kegan Paul, 1982); E. Ann Kaplan, *Women and Film: Both Sides of the Camera* (London: Methuen, 1983); Mary Ann Doane, Patricia Mellencamp, and Linda Williams, eds., *Re-Vision: Essays in Feminist Criticism* (Los Angeles: American Film Institute, 1984).

Again as noted in the essay, an introductory understanding of feminist theory is crucial as background to any feminist work in television studies. For a recent overview of feminist theories as they have developed in France, America, and Britain from 1970 to 1986, see Toril Moi, *Sexual/Textual Politics: Feminist Literary Theory* (London: Methuen, 1986). Key texts for recent feminist approaches to television are Nancy Chodorow, *The Reproduction of Mothering: Psychoanalysis and the Sociology of Gender* (Berkeley: University of California Press, 1978); Julia Kristeva, "Women's Time," trans. Alice Jardine and Harry Blake, *Signs* 7, no. 1 (1981): 13–35 (other relevant essays by Kristeva will shortly be available in a collection edited by Toril Moi, to be published by Basil Blackwell); Luce Irigary, "This Sex Which Is Not One," and "When Two Lips Speak Together," both in *This Sex Which Is Not One*, trans. Catherine Porter and Carolyn Burke (Ithaca, N.Y.: Cornell University Press, 1985).

Given the overall perspective of my essay, the most useful essays on television criticism for students to start out with are those collected in E. Ann Kaplan, ed., *Regarding Television—Critical Approaches: An Anthology*, American Film Institute Monograph Series, vol. 2 (Frederick, Md.: University Publications of America, 1983). The essays by Robert Stam and Jane Feuer provide excellent background for work on the televisual apparatus that is central to any specifically *feminist* analysis, and those by Tania Modleski, Charlotte Brunsdon, Sandy Flitterman, and Robert C. Allen provide models for different feminist approaches to the soap opera.

Essays that provide important background for work in the collection include Stephen Heath and Gillian Skirrow, "Television: A World in Action," *Screen* 18, no. 2 (1977): 7–59; Janice Winship, "Handling Sex," *Media, Culture and Society* 3, no. 1 (1981): 6–18; and, specifically on soaps, Ellen Seiter, "The Role of the Woman Reader: Eco's Narrative Theory and Soap Operas," *Tabloid* 6 (1981): 36–43. For a detailed analysis of soaps and a full bibliography, see Robert C. Allen, *Speaking of Soap Operas* (Chapel Hill: University of North Carolina Press, 1985). Chapters 3 and 4, in particular, contain material relevant to feminist criticism.

Recently, some feminist scholars have begun to extend work being done on melodrama, in relation to the Hollywood film, to television serials. For film background, see the essays collected in Laura Mulvey and Jon Halliday, eds., *Douglas Sirk* (Edinburgh: Edinburgh Film Festival, 1972); Laura Mulvey, "Afterthoughts on 'Visual Pleasure and Narrative Cinema,' Inspired by *Duel in the Sun*," *Framework*, no. 15/16/17 (1981): 12–15; Mary Ann Doane, "The Woman's Film: Possession and Address," in *Re-Vision*, pp. 67–82; E. Ann Kaplan, "Theories of Melodrama," *Women and Performance* 1, no. 1 (Summer 1983): 40–48; and Christine Gledhill, ed., *Women and Melodrama* (London: British Film Institute, 1986). For applications to television, see Jane Feuer, "Melodrama, Serial Form and Television Today," *Screen* 25, no. 1 (January/February 1984): 4–16; and Annette Kuhn, "Women's Genres," *Screen* 25, no. 1 (January/February 1984): 18–28.

Another important area of research is related to discourse theory, rather than to either psychoanalytic or semiological approaches, although the work is linked to both by an interest in the construction of the subject through the processes of desire. Both Fina Bathrick (in a forthcoming book on the media) and Lynn Spigel (in her dissertation at UCLA) deal with discourses that address the relationship between television, the home, and the family in advertising and women's magazines. In two unpublished papers ("Installing the Television Set: The Discourses of Women's Home Magazines, 1948–1955," and "TV in the TV Home: Television's Discourse on Television, 1948–1955") read at the Society for Cinema Studies Conference in April 1986, Spigel looks at where and how TV is represented, as for example in books on interior decor or in TV programs themselves. Spigel shows how the discourses she discovers dramatize the problematic of TV's relationship to the public in their construction of a series of subject positions for family members in the home equipped with TV.

This work on discourse analysis is being taken up in another important area, little dealt with from the cultural-studies theoretical position—namely that of research in audience response to television. The best work here is informed by neo-Marxist Althusserian ideas, sometimes together with Foucaultian theory (for a good example, see Tony Bennett, "Texts in History: The Determinations of Readings and Their Texts," in *Post-Structuralism and the Question of History*, ed. D. Attridge et al. [Cambridge: Cambridge University Press, 1986]). This work is important for feminist criticism because it combines the problematic of subject formation crucial to gender issues with equally central issues of contextual and historical specificities. Continuing their work on the audience for the British program *Nationwide*, David Morley and Charlotte Brunsdon are studying how the television operates within the home setting—looking at the gender discourses that structure its use, and at how these determine who controls the set, and so on. For earlier work, see Brunsdon and Morley, *Everyday Television: "Nationwide"* (London: British Film Institute, 1978).

Feminist criticism of music television is only just starting. The earliest piece is E. Ann Kaplan, "A Postmodern Play of the Signifier? Advertising, Pastiche and Schizophrenia in Music Television," paper delivered in summer 1984 at the London International Television Conference and published in *Television in Transition*, ed. Phillip Drummond and Richard Paterson (London: British Film Institute, 1985), pp. 146–63; a development of these ideas is available in Kaplan, "Sexual Difference, Pleasure and the Construction of the Spectator in Music Television," *Oxford Literary Review* 8, no. 1/2 (1986): 113–23; further work on sexual difference in rock videos may be found in her forthcoming book, *Rocking around the Clock: Consumption and Postmodern Culture in Music Television* (London: Methuen, 1987).

For an example of a quantifying approach to sex roles in music television, see Jane D. Brown and Kenneth Campbell, "The Same Beat but a Different Drummer: Race and Gender in Music Videos," *Journal of Communication* 36, no. 1 (Winter 1986): 94–106; for a useful collection of essays on music television relevant to a feminist analysis, see *The Journal of Communication Inquiry* 10, no. 1 (Winter 1986).

BRITISH CULTURAL STUDIES AND TELEVISION
JOHN FISKE

The term "culture," used in the phrase "cultural studies," is neither aesthetic nor humanist in emphasis, but political. Culture is not conceived of as the aesthetic ideals of form and beauty to be found in great art, nor in more humanist terms as the voice of the "human spirit" that transcends boundaries of time and nation to speak to a hypothetical universal man (the gender is deliberate—women play little or no role in this conception of culture). Culture is not, then, the aesthetic products of the human spirit acting as a bulwark against the tide of grubby industrial materialism and vulgarity, but rather a way of living within an industrial society that encompasses all the meanings of that social experience.

Cultural studies is concerned with the generation and circulation of meanings in industrial societies (the study of culture in nonindustrial societies may well require a different theoretical base, though Levi-Strauss's work has proved of value in studying the culture of both types of society). But the tradition developed in Britain in the 1970s was necessarily focused on culture in industrial societies. In this chapter I shall draw largely upon the work done at the University of Birmingham Centre for Contemporary Cultural Studies (CCCS) under Stuart Hall, with some references to the work of Raymond Williams and of *Screen*. The cultural studies developed at the CCCS is essentially Marxist in the traditions developed by Louis Althusser and Antonio Gramsci, though the Marxism can be inflected sometimes with a structuralist accent, sometimes with an ethnographic one.

Some basic Marxist assumptions underlie all the British work in cultural studies. They start from the belief that meanings and the making of them (which together constitute culture) are indivisibly linked to the social structure and can only be explained in terms of that structure and its history. Correlatively, the social structure is held in place by, among other forces, the meanings that culture produces; as Stuart

Hall says: "A set of social relations obviously requires meanings and frameworks which underpin them and hold them in place."[1] These meanings are not only meanings of social experience, but also meanings of self, that is, constructions of social identity for people living in industrial capitalist societies that enable them to make sense of themselves and of their social relations. Meanings of experience and meanings of the subject (or self) who has that experience are finally part of the same cultural process.

Also underlying this work is the assumption that capitalist societies are divided societies. The primary axis of division was originally thought to be class, though gender may now have replaced it as the most significant producer of social difference. Other axes of division are race, nation, age group, religion, occupation, education, political allegiance, and so on. Society, then, is not an organic whole but a complex network of groups, each with different interests and related to each other in terms of their power relationship with the dominant classes. Social relations are understood in terms of social power, in terms of a structure of domination and subordination that is never static but is always the site of contestation and struggle. Social power is the power to get one's class or group interest served by the social structure as a whole, and social struggle—or in traditional Marxist terms, the class struggle—is the contestation of this power by the subordinate. In the domain of culture, this contestation takes the form of the struggle for meaning, in which the dominant classes attempt to "naturalize" the meanings that serve their interests into the "common sense" of the society as a whole, whereas subordinate classes resist this process in various ways, and to varying degrees, and try to make meanings that serve their interests. The current work of the feminist movement provides a clear example of this cultural struggle and contestation. Angela McRobbie, for instance, shows how young girls are able to contest the patriarchal ideology structured into films such as *Flashdance* and produce feminine readings of them.[2] Annette Kuhn, working in cinema, shows how women can recover feminine discourse that resists the dominant ideology as it works through the structure of the text.[3]

But the attempt of the dominant classes to naturalize their meanings is rarely, if ever, the result of a conscious intention of individual members of those classes, (though resistance to it is often, though not always, conscious and intentional). Rather, it must be understood as the work of ideology inscribed in the cultural and social practices of a class and therefore of the members of that class. And this brings us to another basic assumption: culture is ideological.

The cultural studies tradition does not see ideology in its vulgar Marxist sense of "false consciousness," for that has built into it the assumption that a true consciousness is not only possible, but will actually occur when history brings about a proletarian society. This sort of idealism seems inappropriate to the late twentieth century, which appears to have demonstrated not the inevitable self-destruction of capitalism, but its unpredicted (by Marx) ability to reproduce itself and to incorporate into itself the forces of resistance and opposition. History casts doubt on the possibility of a society without ideology, in which people have a true consciousness of their social relations.

Structuralism, another important influence on British cultural studies, also denies the possibility of a true consciousness, for it argues that reality can only be made sense of through language or other cultural meaning systems. Thus the idea of an objective, empirical "truth" is untenable. Truth must always be understood in terms of how it is made, for whom and at what time is it "true." Consciousness is never the product of truth or reality, but rather of culture, society, and history.

Althusser and Gramsci were the theorists who, between them, offered a way of accommodating both structuralism (and, incidentally, Freudianism) and the history of capitalism in the twentieth century with Marxism. For Althusser, ideology is not a static set of ideas imposed upon the subordinate by the dominant classes but rather a dynamic process that is constantly reproduced and reconstituted in practice—that is, in the way that people think, act, and understand themselves and their relationship to society.[4] He rejects the old idea that the economic base of society determines the whole of the cultural superstructure. He replaces this base/superstructure model with his theory of overdetermination, which not only allows the superstructure to influence the base but also produces a model of the relationship between ideology and culture that is not determined solely by economic relations. At the heart of this theory is the notion of the ideological state apparatuses (ISAs), by which he means social institutions such as the family, the education system, language, the media, the political system, and so on. These institutions produce in people the tendency to behave and think in socially acceptable ways (as opposed to the repressive state apparatuses such as the police force or the law, which coerce people into behaving according to the social norms). The social norms, or that which is socially acceptable, are of course neither neutral nor objective—they have developed in the interests of those with social power, and they work to maintain their sites of power by naturalizing them into the common-sense, the only, social positions for power to be located. Social norms are both ideologically slanted in favor of a particular class or group of classes and accepted as natural by other classes, even when the interests of these classes are directly opposed by the ideology that is reproduced when life is lived according to those norms.

These norms are realized in the day-to-day workings of the ideological state apparatuses. Each one of these institutions is "relatively autonomous," and there are no overt connections between it and any of the others—the legal system is not explicitly connected to the school system nor to the media, for example—yet they all perform similar ideological work. They are all patriachal; they are all concerned with the getting and keeping of wealth and possessions; and they all assert individualism and competition between individuals. But the most significant feature of ISAs is that they all present themselves as socially neutral, as not favoring one particular class over any other. Each presents itself as a principled institutionalization of equality—the law, the media, and education all claim loudly and often to treat all individuals equally and fairly. The fact that the norms that are used to define equality and fairness are those derived from the interests of the white, male, middle classes is more or less adequately disguised by these claims of principle, though feminists and those working for racial and class harmony may claim that this disguise can be torn off relatively easily.

Althusser's theory of overdetermination explains this congruence between the "relatively autonomous" institutions not through their relationship to a common, determining economic base, but to an overdetermining network of ideological interrelationships among all of them. The institutions appear autonomous only at the official level of stated policy, though the belief in this "autonomy" is essential for their ideological work. At the unstated level of ideology, however, each institution is related to all the others by an unspoken web of ideological interconnections, so that the operation of any one of them is "overdetermined" by its complex, invisible network of interrelationships with all the others. Thus, the education system, for example, cannot tell a different story about the nature of the individual from that told by the legal system, the political system, the family, and so on.

Ideology is not, then, a static set of ideas through which we view the world, but a dynamic social practice, constantly in process, constantly reproducing itself in the ordinary workings of these apparatuses. But it works at this macro-institutional level only because it works similarly at the micro-level of the individual. To understand this we need to replace the idea of the individual with that of the subject. The individual is produced by nature, the subject by culture. Theories of the individual concentrate on differences between people and explain these differences as natural. Theories of the subject, on the other hand, concentrate on people's common experiences in a society as being the most productive way of explaining who (we think) we are. Althusser believes that we are all constituted as subjects-in-ideology by the ISAs, that the ideological norms naturalized in their practices not only constitute the sense of the world for us, but they also constitute our sense of ourselves, our sense of identity, and our sense of our relations to other people and to society in general. Thus, we are each of us constituted as a subject in, and subject to, ideology. The subject, therefore, is a social construction, not a natural one. Thus a biological female can have a masculine subjectivity (that is, she can make sense of the world and of her self and her place in that world through patriarchal ideology). Similarly, a black can have a white subjectivity and a member of the working classes a middle-class one.

The media and language play an important part in this constant construction of the subject, by which we mean the constant reproduction of ideology in people. Althusser uses the words "interpellation" and "hailing" to describe this work of the media. These terms derive from the idea that any language, whether it be verbal, visual, tactile, or whatever, is part of social relations, and that in communicating with someone we are reproducing social relationships.

In communication with someone, our first job is to "hail" them, almost as if hailing a cab. To answer, they have to recognize that it is them, and not someone else, that we are talking to. This recognition derives from signs, carried in our language, of whom we think they are. We will hail a child differently from an adult, a male differently from a female, someone whose status is lower than ours differently from someone in a higher social position. In responding to our hail, the addressee recognizes the social position our language has constructed, and if their response is cooperative, they adopt this position. Hailing is the process by which language identifies and constructs a social position for the addressee. Interpellation is the larger process whereby language constructs social relations for both parties in an act of communication and thus locates them in the broader map of social relations in general.

For example, the television series *The A-Team* interpellates the viewer as masculine, as desiring power, and as an individual within a team. In accepting the role of addressee of the program, we are not only adopting the subject position interpellation has produced for us, but we are, by the act of viewing cooperatively, reproducing the ideology of masculinity in the social practice of watching television and in ourselves as subjects in patriarchy. The program hails *us,* and in recognizing ourselves as the *we* being spoken to, we are constituting ourselves-as-subjects in the ideological definition of *us* that the program proposes.

As Mimi White points out in her chapter in this volume, this view of ideology as a process constantly at work, constructing people as subjects in an ideology that always serves the interests of the dominant classes, found powerful theoretical support in Gramsci's theory of hegemony. Originally, hegemony was used to refer to the way that a nation could exert ideological and social, rather than military or coercive, power over another. However, cultural theorists tend to use it to describe the process by which a dominant class wins the willing consent of the subordinate classes to the system that ensures their subordination. This consent must be constantly won and rewon, for people's material social experience constantly reminds them of the disadvantages of subordination and thus poses a constant threat to the dominant. So, like Althusser's theory of ideology, hegemony is not a static power relationship, but a constant process of struggle in which the big guns are on the side of those with social power, but in which victory does not necessarily go to the big guns—or, at least, in which victory is not necessarily total. Indeed, the theory of hegemony foregrounds the notion of ideological struggle much more than does Althusser's ideological theory, which at times tends to imply that the power of ideology and the ISAs to form the subject in ways that suit the interests of the dominant class is almost irresistible. Hegemony, on the other hand, posits a constant contradiction between ideology and the social experience of the subordinate that makes this interface into an inevitable site of ideological struggle. In hegemonic theory, ideology is constantly up against forces of resistance. Consequently it is engaged in a constant struggle not just to extend its power, but even to hold on to the territory it has already colonized.

This definition of culture as a constant site of struggle between those with and those without power underpins the most interesting current work in cultural studies. Earlier work in the tradition tended to show how the dominant ideology reproduced itself invisibly and inevitably in the forms of popular television.[5] Hall's influential essay "Decoding and Encoding" is often seen as a turning point in British cultural studies, for it opened up the idea that television programs do not have a single meaning, but are relatively open texts, capable of being read in different ways by different people. Hall also suggests that there is a necessary correlation between people's social situations and the meanings that they may generate from a television program. He thus postulates a possible tension between the structure of the text, which necessarily bears the dominant ideology, and the social situations of the viewers, which may position them at odds with that ideology. Reading or viewing television, then, becomes a process of negotiation between the viewer and the text. Use of the word "negotiation" is significant, for it implies both that there is a conflict of interests that needs to be reconciled in some way, and that the process of reading television is one in which the reader is an active maker of meanings from the text, not a passive recipient of already constructed ones.

Hall developed his theory of the "preferred reading" to account for this conflict of interests.[6] He postulates three broad reading strategies that are produced by three generalized, not material, social positions that people may occupy in relation to the dominant ideology. These are the dominant, the negotiated, and the oppositional. The dominant reading is produced by a viewer situated to agree with and accept the dominant ideology and the subjectivity that it produces. A negotiated reading is one produced by a viewer who fits into the dominant ideology in general, but who needs to inflect it locally to take account of his or her social position. This inflection may contain elements of resistance deriving from the perception of areas of conflict between the constructions of the dominant ideology and the viewer's more materially based construction of social experience. And finally, there are readings produced by those whose social situation puts them into direct opposition with the dominant ideology—these readings are termed oppositional.

The preferred reading theory proposes that TV programs generally prefer a set of meanings that work to maintain the dominant ideologies, but that these meanings cannot be imposed, only preferred. Readers whose social situations lead them to reject all or some constructions of the dominant ideology will necessarily bring this social orientation to their reading of the program.

An example may make these ideas clearer and may also show up some problems associated with them. *Magnum, P.I.* has been a popular TV private-eye show since it first screened in December 1980. A dominant reader of the show would find pleasure in it because it reproduces in him/her a subject position that fits easily into the dominant ideology, that demonstrates that ideology as an adequate way of making sense of the world, and that therefore affirms the subject position as the natural one from which to view the world. Magnum, played by Tom Selleck, can be seen as literally embodying patriarchal capitalism. The dominant reader, therefore, whose enjoyment of the program consists in identifying with the hero, is hailed as a subject in patriarchal capitalist ideology. The ideology works both through the progress and resolution of each week's narrative, and through the frame of that narrative—that is, those elements of the program that are consistent from week to week. They are not part of the conflict to be resolved in each episode and therefore form the basic, uninspected assumptions, or common sense, through and in which the dominant ideology naturalizes itself. This narrative frame is crucially manifest in the main characters: the hero and his two helpers, Higgins and T.C., who together constitute the hero-team. The dominant ideology works in a number of overlapping specific ideologies: the ideologies of masculinity, of individualism, of competition, and so on. As our analysis proceeds, we will see how impossible it is to separate these specific ideologies from each other; they all merge "naturally" into the general ideology of patriarchal capitalism.

Magnum and his team are male and masculine. Maleness is a fact of nature, but masculinity is a cultural constraint that gives meaning to maleness, by opposing it to femininity. Shere Hite investigated men's opinions of what makes a man a man. The list of characteristics she generated began with such qualities as self-assurance, lack of fear, the ability to take control, autonomy and self-sufficiency, leadership, dependability, and achievement. These qualities work along two main dimensions: self-sufficiency, which stresses the absence of a need to depend on others; and assertiveness, expressed as the ability to lead others and to influence events, and most readily experienced in performance and achievement.[7] Freudian explanations of how masculinity is achieved in childhood point to the boy's rejection of his desire for his mother because of the position of rivalry with his father that it puts him into. He then identifies with his father in order to gain access to masculine power and authority. The price he pays, however, is the guilt-producing rejection of his mother and the consequent suppression in himself of the feminine characteristics that threaten this male power and independence. These characteristics are essentially ones of nurturing and of intimacy. The absence of women from significant roles in *Magnum, P.I.* is a representation of the suppression and devaluation of feminine characteristics in patriarchal constructions of masculinity.

Like all ideological constructs, masculinity is constantly under threat—it can never rest on its laurels. The threats come internally from its insecure bases in the rejection of the mother (and the guilt that this inspires) and the suppression of the feminine, and externally from social forces, which may vary from the rise of the women's movement to the way that the organization of work denies many men the independence and power that their masculinity requires. Thus masculinity has to be constantly reachieved, rewon. This constant need to reachieve masculinity is one of the underlying reasons for the popularity of the frequent televisual display of male performance. There is a link here between TV programs like *Magnum, P.I., The A-Team, Knight Rider, T. J. Hooker* (to name but a few) and pornography. For, as Andrew Moye points out, pornography reduces masculinity to performance, in this case, the performance of the penis.[8] Masculinity must, in patriarchy, be able to cope with any situation; it becomes less a construction of man than of superman. There is always a gap between the actual male performance and the supermale performance proposed by patriarchy that these programs are striving to close. Similarly, there is always a gap between the penis and the phallus that pornography strives to close. The penis is the natural sign of maleness; the phallus is the cultural sign of masculinity—the totality of meanings, rights, and power that a culture ascribes to maleness. Hence, these shows, in their role as "masculine definers," are full of phallic symbols, particularly guns as agents of male power (think how rare it is for a female on TV to use a gun successfully, particularly to kill a male).

This male power must be tempered with notions of duty and service—it must be used in the interest of the weak or of the nation. If used for personal gain it becomes the mark of the villain. So masculine power involves both exerting and submitting to authority. This is one of the reasons why the male team or duo is such a popular formation of the masculine hero, and why this hero formation so commonly works on the side of, but in tension with, an institution of official authority. Another reason is that the male bonding inherent in such a formation allows for an intimacy that excludes the threat of the feminine. Feminine intimacy centers on the relationship itself and produces a dependence on the other that threatens masculine independence—consequently, any woman who attracts a hero has to be rejected at the end of the episode. Male bonding, on the other hand, allows an interpersonal dependency that is goal-centered, not relationship-centered, and thus serves masculine performance instead of threatening it. The hero team also compensates for male insecurity: any inadequacies of one member are compensated by the strengths of another, so the teams become composite constructions of masculinity. All the traits embodied in one man would make him into a unbelievable superman, and ideology—closely connected to fantasy though it be—has to be grounded in credibility, that is, in a conventional construction of the realistic. If it were not it would be unable to work on, and be put to work by, the viewers.

I have concentrated on how the ideology of masculinity is actively working in the popularity of *Magnum, P.I.* and its ilk. It is comparatively easy to see how this merges indistinguishably into the overlapping ideologies of individualism, of competition, and of a form of "social Darwinism" that proposes that morality is always on the side of eventual winners. These ideologies, in turn, merge into a particular construction of American and Western nationalism—a Reaganite version of the nation that sees it as masculine (exerting in the international sphere power over others in the service of the weak or of a higher morality), based on competitive individualism and social Darwinism. The fact that Magnum's "masculinity" was developed in Vietnam (as was that of the A-Team, T. J. Hooker, one of the Simon brothers) is not only part of the American rehabilitation of Vietnam (and of Thatcher's Falkland actions, Reagan's Achille Lauro "rescue," Israel's Entebbe) but also works ideologically to ground problematic political acts in the much less questioned and therefore more natural-seeming construction of masculinity.

Magnum's hero team is constructed to embody, not just the ideology of masculinity, but also the overlapping ideologies of race and nation. In the domain of masculinity, Higgins, the majordomo, represents service and submission to authority, whereas T. C. the driver/pilot and engineering expert, represents masculinity as physical power and its mechanical extensions. His blackness (like that of B.A. in *The A-Team,* who performs a similar ideological role) introduces the racial dimension—physical power may be the basis of masculinity, but because it needs leadership and social control to be acceptable, it is therefore low in the hierarchy of masculine traits. It is noticeable how often the hero team contains a nonwhite in a subordinate position, from Ahab and Queequeg in *Moby-Dick,* through the Lone Ranger and Tonto, to the television hero teams of *Ironside, The A-Team*, and *Magnum, P.I.* (In *Starsky and Hutch*, Starsky, the dark Jewish one, is the driver; Hutch, the blond Aryan, is college educated and the leader. Their superior officer is a black, but, as is often the case, the role of the official superior is narratively subordinate to the hero team.) Higgins in *Magnum, P.I.* is English, embodying variously the meanings of tradition, of the old-fashioned, of weakness, of excessive submission to authority and convention. These traits can be used for comic effect, but can just as frequently be used as ideological underwriters of the more aggressive, pragmatic, even amoral American values embodied in Magnum himself. The ideologies of masculinity, race, and nation enter into a mutually guaranteeing relationship in the formation of the hero team.

The reader whose social position is one of ease with the dominant ideology, who works *with* the program, will use its foregrounded ideology to reaffirm his (gender deliberate) ideological frame, through which he views the world and makes sense of both himself and his social experience. In responding to the program's interpellation, he adopts the subject position it constructs for him. Althusser's account of the power of the dominant ideology working through language and texts to construct the reader as a subject in ideology can really only account for Hall's "dominant reading." Gramsci's notion of hegemony, with its emphasis on the dominant ideology's constant struggle to win the consent of the subordinate and to incorporate or defuse oppositional forces, underlies Hall's next two reading strategies—those that produce negotiated and oppositional readings.

A negotiated reading is one that inflects the dominant ideology towards the social experience of a particular viewing group. Thus, boys watching *Magnum, P.I.* might concentrate on the performance side.

Their social situation denies them the ability to exert the power (either physically, because their bodies are still immature, or socially, because of their low hierarchical position in the family or school) that society tells them they should if they are to be "masculine." They may thus identify with the mechanical expertise of T. C. and "negotiate" a more important role for him, finding in his mechanical power an essential element of Magnum's success. We know that B.A., the muscular black driver and mechanic in *The A-Team,* is particularly popular with white youths. Presumably they foreground his strength and engineering expertise over his race and therefore make sense of his subordinate position in the team as a way of articulating their subordination in society, not the powerlessness of blacks in a white hegemony. Black youths, however, would be more likely to use his blackness, his strength, and the gold chains he always wears (which Mr. T says are symbols of his people's slavery) to make sense of their constant struggle to assert and extend their own position in society. Similarly, older men working in corporate America might find in Higgins values that enable them to relate Magnum's independence to their own positions within a constraining work organization. Other men who feel that the responsibilities of the family contradict their masculine needs for independence and freedom might see the frequent arguments between Magnum and Higgins as an expression of this. They might ignore Higgins's Englishness or his occasional role of bringing the validation of history and tradition to contemporary American assertiveness.

Female viewers of the program will also negotiate it toward their interests. The physical attractiveness of Magnum may be read as an integral part of his protection of the weak. His rejection of intimacy with any one woman would not be seen as a latent recognition of women's threat to masculinity, or as a representation of the suppression of the feminine in the masculine psyche and therefore of the subordination of women in a patriarchal society (for the two are structural reflections of each other). It would rather be seen as a means of maintaining his masculine freedom to serve all women and provide them with the security and justice that their material social position may deny them. Magnum can be read, then, not as the embodiment of masculine oppression in patriarchy, but as the patriarchal agent that rights the wrongs and corrects the deficiencies of the system in practice.

These sorts of negotiated readings are ones produced by the ideologically cooperative readers—they read "with" the structure of the text and seek to match their social experience with the ideology-in-the-text. Actually they produce almost dominant readings, which may lead us to speculate whether the "pure" dominant reading is ever achieved. There is probably no one audience group positioned in perfect ideological centrality. All groups will need to "shift" the text slightly to fit their social position. In this case, all readings become, as Horace Newcomb suggests, negotiated ones.[9] But if this is so, it is still valuable to recognize that negotiated readings can occur on a scale stretching from the ideologically central to the deviant. Thus, a macho teenager, at the point of maximum opposition to authority, may read the violence in Magnum as justified masculinity that overrides the "weakness" of its use in the service of the weak or of "natural justice." Such a reading may see the failure of the police or official authorities as a criticism of them and of the society they stand for. Magnum and his team can be read as masculine assertiveness operating outside and in opposition to the forces of law and order. Such a reading veers toward the oppositional because it plays down the contextual ideologies within which that of masculinity operates and from which it acquires its social and moral acceptability.

Readings at this end of the scale stop being negotiated and become oppositional when they go "against" the text to deconstruct the dominant ideology. Thus, a feminist could read *Magnum, P.I.* as a blatant display of patriarchal chauvinism and how it sells itself to society—Magnum as male hegemony. This reading would produce, not pleasure (except the wry pleasure of recognizing patriarchy up to its tricks yet again), but annoyance. The annoyance could be used to activate political action, whether in the sphere of consciousness-raising or more directly. Similarly, a black activist could find the subordinate position of T. C. in the hero formation a perfect example of white hegemony at work and a spur to further oppositional practice. But such oppositional readers would be unlikely to watch the program regularly and thus would form a minute fraction of its total audience.

The typical reading, therefore, is likely to be, as Newcomb agrees, a negotiated one.[10] This would appear to be a theoretical truism of the cultural studies approach, though it is one that is not often made explicitly. For if society is seen not as homogenous, but as a structure of different interest groups, and if television is to appeal to a large number of people in a society, then it follows that the television audience must not be seen as a homogenous mass, but as a mix of social groups each in a different relationship to the dominant ideology. However complex and difficult it might be to describe these relationships, they can always be placed on a scale that ranges from *acceptance of* to *opposition to* the dominant ideology. The television text can only be popular if it is open enough to admit a range of negotiated readings, through which various social groups can find meaningful articulations of their own relationship to the dominant ideology. Any television text must, then, be polysemic for the heterogeneity of the audience requires a corresponding heterogeneity of meanings in the text. The hero-team is a significant ideological formation here, as it provides for a greater "openness" than the single hero. Its greater variety of opportunities for identification enables various social groups to negotiate appropriate points of entry into the dominant ideology.

An important body of cultural studies work has derived from the recognition of the polysemic nature of TV texts and the heterogeneity of audiences, a strand that we might label the "ethnographic." Workers such as David Morley, Dorothy Hobson, Angela McRobbie, and Robert Hodge and David Tripp have set out to discover how actual audience groups actively use television as part of their own culture—that is, use it to make meanings that are useful to them in making sense of their own social experience and therefore of themselves.[11] In this, these scholars are in opposition to the other dominant strand of British (and European) study of culture, which is that centered around *Screen* and has come to be known as Screen Theory. Screen Theory draws upon a combination of structuralism and semiotics with psychoanalysis and Marxism to argue the power of the text over the viewing subject and to analyze, with great theoretical sophistication, the textual strategies that operate to position the viewing subject within dominant ideology. Morley clearly elaborates the theoretical and methodological differences between the two schools.[12]

Morley tested Hall's preferred reading theory in the field. He took a television program that he and Charlotte Brunsdon had previously subjected to detailed cultural analysis, showed it to groups of people, and then held discussions on their reactions to it and its meanings for them.[13] He turned to groups rather than individuals because he was interested in the shared, and therefore social, dimensions of reading. The groups were defined largely by occupation—bank managers, apprentices, students, trade unionists, and so on—because occupation is a prime definer of social class, and class was, in Hall's theory, the prime producer of social difference and therefore of different readings. (A few of Morley's groups, however, were defined by gender or race— black unemployed women, for example.) What Morley found was that Hall had overemphasized the role of class in the production of semiotic differences and had underestimated the variety of readings that could be made. Thus the readings showed some interesting and unexpected cross-class similarities: bank managers and apprentices, for example, produced broadly similar readings despite their class differences; so, too, did some university students and shop stewards. We could explain these apparent anomalies by suggesting that the apprentices and bank managers were similarly constructed as subjects of a capitalist ideology, in that both were inserting themselves into the dominant system (albeit at different points) and thus had a shared interest in its survival and success. Some university students (not all, by any means) and trade union officials, however, were in institutions that provided them with ways of criticizing the dominant system and they thus produced more oppositional readings.

Morley's work showed that Hall's three categories of reading were too simplistic. The wide range of readings resisted simple categorization based primarily on the single factor of class, so he replaced Hall's model with one derived from discourse theory. A discourse is a socially produced way of talking or thinking about a topic. It is defined by reference to the area of social experience that it makes sense of, to the social location from which that sense is made, and to the linguistic or signifying system by which that sense is both made and circulated. When the media report, as they typically do, that management "offer," but trade unions "demand," they are using the mass media discourse of industrial relations, which is located in a middle-class position. They could equally well report (but never do) that the unions "offered" to work for an extra 5 percent, but management "demanded" that they worked for 2 percent. The consistent ascription of the generous "offer" and the grasping "demand" to management and unions, respectively, is clear evidence of the social location of this particular discourse. A discourse, then, is a socially located way of making sense of an important area of social experience.

A television text is, therefore, a discourse (or a number of discourses if it contains contradictions), and the reader's consciousness is similarly made up of a number of discourses through which s/he makes sense of his/her social experience. Morley defines reading a television text as that moment when the discourses of the reader meet the discourses of the text. Reading becomes a negotiation between the social sense inscribed in the program and the meanings of social experience made by its wide variety of viewers; this negotiation is a discursive one.

Cultural studies sees the television experience (that is, the entity constituted by the text and the activity of viewing it) as a constant dynamic movement between similarity and difference. The dimension of similarity is that of the dominant ideology that is structured into the forms of the program and is common to all the viewers for whom that program is popular. The dimension of difference, however, accounts for the wide variety of groups who must be reached if the program is to be popular with a large audience. These groups will align themselves to the dominant ideology in different ways, and these ways will be paralleled in the different readings they make of the program that is common, or similar, to all. The play between similarity and difference is one way of experiencing the struggle between hegemony and resistance.

This emphasis on discourse and the reader necessarily reduces the prime position granted to the text in the cultural theory of the 1970s. The text can no longer be seen as a self-sufficient entity that bears its own meaning and exerts a similar influence on all its readers. Rather, it is seen as a potential of meanings that can be activated in a number of ways. Of course, this potential is proscribed and not infinite; the text does not determine its meaning so much as delimit the arena of the struggle for that meaning by marking the terrain within which its variety of readings can be negotiated. This discursive negotiation that we now understand reading to be also means that the boundaries of the text are fluid and unstable. Raymond Williams suggested in the early seventies that television was not a discrete series of programs or texts but "a flow" in which programs, commercials, newsbreaks, and promos all merged into a continuous cultural experience. More recently, John Hartley has suggested that television is a "leaky" medium whose meanings constantly spill over into other areas of life.[14]

Angela McRobbie has also explored the permeability of the boundary between television and other forms of cultural experience. Her study of girls and dance finds similar pleasures and meanings in girls' dancing in discos and in their viewing of films such as *Flashdance* or films and television programs such as *Fame*.[15] On one level of reading, the narrative form and pleasure of *Flashdance* clearly work hegemonically—the female factory worker uses her skill in breakdancing to win a place in a ballet company and marry the boss's son. In the process she displays her body for patriarchal pleasure; indeed, her beautiful body is crucial to her successful move up the social hierarchy (dancing ballet, rather than break, and marrying into management). Women, so the hegemonic reading would go, are rewarded for their ability to use their beauty and talents to give pleasure to men. But McRobbie has shown that this is not the only reading. She has found among teenage girls a set of meanings for dance and female sexuality that contest and struggle against the patriarchal hegemony. For these girls, dance is a form of autoeroticism, a pleasure in their own bodies and sexuality that gives them an identity not dependent upon the male gaze of approval. *Their* discourse of dance gives a coherent meaning to dancing in discos or to watching filmic and televisual representations of dance that assert their subcultural identity and difference from the rest of society. Its meanings for them were meanings that they had made out of the forms that had been provided for them by patriarchy. Exploring the strategies by which subordinate subcultures make their own meanings in resistance to the dominant is currently one of the most productive strands in cultural studies.

Madonna, who, in 1986 at least, has been a major phenomenon of popular culture, can provide us with a good case study. Her success has arguably been due largely to television and to her music videos; most critics have nothing good to say about her music, but they have a lot to say about her image—"the Madonna look." In the autumn of 1984 she was signed to Sire Records, which is "where Warner Brothers put people they don't think will sell."[16] She got some dance club play for "Borderline" and "Holiday," but the *Madonna* LP was selling only slowly. *Like a Virgin,* her second LP, had been made but not released. Warner Brothers then gave Arthur Pierson a tiny budget to make a rock video of "Lucky Star." He shot the video in an afternoon, against a white studio backdrop, and it pushed the song into the top ten. The *Madonna* album's sales followed suit, and *Like a Virgin* was released for the Christmas market. Both LPs held the number-one position for a number of weeks. The film *Desperately Seeking Susan* was released in March 1985 and added an adult audience to the teenage (largely female) one for the songs and videos. The film supported the videos in establishing the "Madonna look," a phrase repeated endlessly by the media in 1985 capitalized on by Madonna when she established her *Boy Toy* label to sell crucifix earrings; fingerless lace gloves; short, navel-exposing blouses; black, lacy garments; and all the other visual symbols she had made her own. A concert tour started in April (in the foyers, of course, Madonna-look items were for sale) and an old film—*A Certain Sacrifice*—that never made cinema release was dug up for the home-video market. Also dug up and published in *Playboy* and *Penthouse* were old art-school nude photos, and, at the end of 1985, her marriage to Sean Penn became a worldwide multimedia event despite its "secret" location. In other words, Madonna is a fine example of the capitalist pop industry at work, creating a (possibly short-lived) fashion, exploiting it to the full, and making a lot of money from one of the most powerless and exploitable sections of the community—young girls.

But such an account is inadequate (though not necessarily inaccurate as far as it goes) because it assumes that the Madonna fans are, in Stuart Hall's phrase, "cultural dupes," able to be manipulated at will and against their own interests by the moguls of the culture industry.[17] Such a manipulation is not only economic, but also ideological, because the economic system requires the ideology of patriarchal capitalism to underpin and naturalize it; economics and ideology can never be separated. There is plenty of evidence to support this view, too. Madonna's videos exploit the sexuality of her face and body and frequently show her in postures of submission ("Burning Up") or subordination to men. As Ann Kaplan points out in her chapter, her physical similarity to Marilyn Monroe is stressed (particularly in the video of "Material Girl"), an intertextual reference to another star commonly throught to owe her success to her ability to embody masculine fantasy. In the *Countdown* 1985 poll of the top twenty "Sex/Lust Objects," Madonna took third place and was the only female among nineteen males.[18] All this would suggest that she is teaching her young female fans to see themselves as men would see them—that is, she is hailing them as feminine subjects within patriarchy, and as such is an agent of patriarchal hegemony.

But, if her fans are not "cultural dupes," but actively choose to watch, listen to, and imitate her rather than anyone else, there must be some gaps or spaces in her image that escape ideological control and allow her audiences to make meanings that connect with *their* social experience. For many of her audiences, this social experience is one of powerlessness and subordination, and if Madonna as a site of meaning is not to naturalize this, she must offer opportunities for resisting it. Her image becomes, then, not a model meaning for young girls in patriarchy, but a site of semiotic struggle between the forces of patriarchal control and feminine resistance, of capitalism and the subordinate, of the adult and the young.

Cultural studies, in its current state of development, offers two overlapping methodological strategies that need to be combined, and the differences between them submerged, if we are to understand this cultural struggle. One derives from ethnography and requires us to study the meanings that the fans of Madonna actually *do* (or appear to) make of her. This involves listening to them, reading the letters they write to fan magazines, or observing their behavior at home or in public. The fans' words or behavior are not, of course, empirical facts that speak for themselves; they are, rather, texts that need "reading" theoretically in just the same way as the "texts of Madonna" do.

The other strategy derives from semiotic and structuralist textual analysis. This strategy involves a close reading of the signifiers of the text—that is, its physical presence—but recognizes that the signifieds exist not in the text itself, but extratextually, in the myths, countermyths, and ideology of their culture.[19] It recognizes that the distribution of power in society is paralleled by the distribution of meanings in texts, and that struggles for social power are paralleled by semiotic struggles for meanings. Every text and every reading has a social and therefore political dimension, which is to be found partly in the structure of the text itself and partly in the relation of the reading subject to that text.

It follows that the theory informing any analysis also has a social dimension that is a necessary part of the "meanings" that the analysis reveals. Meanings, therefore, are relative and varied. What is constant is the ways in which texts relate to the social system. A cultural analysis, then, will reveal both the way that the dominant ideology is structured into the text and into the reading subject, and those textual features that enable negotiated, resisting, or oppositional readings to be made. Cultural analysis reaches a satisfactory conclusion when the ethnographic studies of the historically and socially located meanings that *are* made are related to the semiotic analysis of the text. Semiotics relates the structure of the text to the social system to explore *how* such meanings are made and the part they play within the cultural process that relates meanings both to social experience and to the social system in general.

So Lucy, a fourteen-year-old fan, says of a Madonna poster: "She's tarty and seductive . . . but it looks alright when she does it, you know, what I mean, if anyone else did it it would look right tarty, a right tart you know, but with her it's OK, it's acceptable. . . . With anyone else it would be absolutely outrageous, it sounds silly, but it's OK with her, you know what I mean."[20] We can note a number of points here. Lucy can find only patriarchal words to describe Madonna's sexuality—"tarty" and "seductive"—but she struggles against the patriarchy inscribed in them. At the same time she struggles against the patriarchy inscribed in her own subjectivity. The opposition between "acceptable" and "absolutely outrageous" refers not only to representations of female sexuality, but is also an externalization of the tension felt by adolescent girls trying to come to terms with the contradictions between a positive feminine view of their sexuality and the alien patriarchal one that appears to be the only one offered by the available linguistic and symbolic systems. Madonna's "tarty" sexuality is "acceptable"—but to whom? Certainly to Lucy, and to girls like her who are experiencing the problems of establishing a satisfactory sexual identity within an oppressing ideology, but we need further evidence to support this tentative conclusion. Matthew, aged fifteen and not a particular fan of Madonna, commented on her marriage in the same discussion. He thought it would last only one or two years, and he wouldn't like to be married to her "because she'd give any guy a hard time." Lucy agreed that Madonna's marriage would not last long but found it difficult to say why except that "Marriage didn't seem to suit her," even though Lucy quoted approvingly Madonna's desire to make it an "open marriage." Lucy's problems probably stem from her recognition that marriage is a patriarchal institution and, as such, is threatened by Madonna's sexuality. The threat of course is not the traditional and easily contained one of woman as whore, but the more radical one of woman as independent of masculinity. As we shall see later, Madonna denies or mocks a masculine reading of patriarchy's conventions for representing women. This might well be why, according to *Time*, many boys find her sexiness difficult to handle and "suspect that they are being kidded."[21] Lucy and Matthew both recognize, in different ways and from different social positions, that Madonna's sexuality can offer a challenge or a threat to dominant definitions of feminity and masculinity.

"Madonna's Best Friend," writing to *Countdown*, also recognizes Madonna's resistance to patriarchy:

I'm writing to complain about all the people who write in and say what a tart and a slut Madonna is because she talks openly about sex and she shows her belly button and she's not ashamed to say she thinks she's pretty. Well I admire her and I think she has a lot of courage just to be herself. All you girls out there! Do you think you have nice eyes or pretty hair or a nice figure? Do you ever talk about boys or sex with friends? Do you wear a bikini? Well according to you, you're a slut and a tart!! So have you judged Madonna fairly?

—Madonna's Best Friend, Wahroonga, New South Wales[22]

This praise for Madonna's "courage just to be herself" is further evidence of the difficulty girls feel in finding a sexual identity that appears to be formed in their interests, rather than in those of the dominant male. Madonna offers some young girls the opportunity to find meanings of their own feminine sexuality that suit them, meanings that are "independent." Here are some other Madonna fans talking: "She's sexy but she doesn't need men. . . . She's kind of there all by herself"; or "She gives us ideas. It's really women's lib, not being afraid of what guys think."[23]

But, like all pop stars, she has her "haters" as well as her fans: "When I sit down on a Saturday and Sunday night I always hear the word Madonna and it makes me sick, all she's worried about is her bloody looks. She must spend hours putting on that stuff and why does she always show her belly button? We all know she's got one. My whole family thinks she's pathetic and that she loves herself.—Paul Young's sexy sneakers."[24] Here again, the "hate" centers on her sexuality and—expressed as her presenting herself in whorelike terms—her painting and displaying herself to arouse the baser side of man. But the sting comes in the last sentence, when the writer recognizes Madonna's apparent enjoyment of her own sexuality, which he (the letter is clearly from a masculine subject, if not an actual male) ascribes to egocentricity and thus condemns.

Madonna's love of herself, however, is not seen as selfish and egocentric by girls; rather, it is the root of her appeal, whose significance becomes clear in the context of much of the rest of the media addressed to them. McRobbie has shown how the "teenage press" typically constructs the girl's body, and therefore her sexuality, as a series of problems—breasts the wrong size or shape, spotty skin, lifeless hair, fatty thighs, problem periods. The list is endless, of course, and the advertisers, the ones who really benefit from these magazines, always have a product that can—at a price—solve the problem.

The polarization of Madonna's audience can be seen in the 1986 *Countdown* polls. She was top female vocalist by a mile (polling four times as many votes as the second choice) and was the only female in the top twenty "Sex/Lust Objects," in which she ranked third. But she was also voted into second place in the "Turkey of the Year" award. She is much loved or much hated, a not-untypical position for woman to occupy in patriarchy, whose inability to understand women in feminine terms is evidenced by the way it polarizes feminity into the opposing concepts of Virgin-Angel and Whore-Devil.

Madonna consciously and parodically exploits these contradictions: "When I was tiny," she recalls, "my grandmother used to beg me not to go with men, to love Jesus and be a good girl. I grew up with two images of women: the virgin and the whore. It was a little scary." She consistently refers to these contradictory meanings of woman in patriarchy. Her video of "Like a Virgin" alternates the white dress of Madonna the bride with the black slinky garb of Madonna the singer; the name Madonna (the virgin mother) is borne by a sexually active female; the crucifixes adopted from nuns' habits are worn on a barely concealed bosom or in a sexually gyrating navel. "Growing up I thought nuns were very beautiful. . . . They never wore any make-up and they just had these really serene faces. Nuns are sexy."[25]

But the effect of working these opposite meanings into her texts is not just to call attention to their role in male hegemony—woman may either be worshipped and adored by man, or used and despised by him, but she has meaning only from a masculine subject position. Rather, Madonna calls into question the validity of these binary oppositions as a way of conceptualizing woman. Her use of religious iconography is neither religious nor sacriligious. She intends to free it from this ideological opposition and to enjoy it, use it, for the meanings and pleasure that it has for *her* and not for those of the dominant ideology and its simplistic binary thinking:

> I have always carried around a few rosaries with me. One day I decided to wear [one] as a necklace. Everything I do is sort of tongue in cheek. It's a strange blend—a beautiful sort of symbolism, the idea of someone suffering, which is what Jesus Christ

on a crucifix stands for, and then not taking it seriously. Seeing it as an icon with no religiousness attached. It isn't sacreligious for me.[26]

The crucifix is neither religious nor sacriligious, but beautiful: "When I went to Catholic schools I thought the huge crucifixes nuns wore were really beautiful." In the same way, her adolescent girl fans find in Madonna meanings of feminity that have broken free from the ideological binary opposition of virgin/whore. They find in her image positive feminine-centered representations of sexuality that are expressed in their constant references to her independence, her being herself. This apparently independent, self-defining sexuality is only as significant as it is because it is working within and against a patriarchal ideology.

In the video "Material Girl," Madonna parodies the whore. Dressed in pink and made up to resemble Marilyn Monroe, she goes through a dance routine with tuxedo-clad young men, using movements and a studio set that are heavy with references to classical Hollywood musicals. During the number she collects jewelry from the men as she sings the refrain, "Cause we're living in a material world, and I am a material girl." But despite her whorelike gathering of riches from men and her singing that only boys with money have any chance with her, she is visually worshipped by the camera and by the choreography—she is the Madonna simultaneously with the whore, however foregrounded the latter may be. It is performances like this that enable *Playboy* to say: "Best of all her onstage contortions and Boy Toy voice have put sopping sex where it belongs—front and center in the limelight."[27]

But the stage performance is embedded in a mininarrative. A poor, sensitive man sees her arrive at the studio, watches her performance, presents her with a simple bunch of daisies in her dressing room afterwards, and drives off with her in an old workman's truck, in which they make love in a rainstorm. The material girl has fallen for the nonmaterial values of love after all. The undermining of the song by the mininarrative may not seem to offer much of a resistance; after all, the main narrative is a conventional romance in which the poor, sensitive man is finally preferred to the apparently more attractive, rich one. The "true love" that triumphs is as much a part of patriarchal capitalism as the materialism it defeats. But this contradiction does not work alone—it is supported by parody, by puns, and by Madonna's awareness of *how* she is making an image, not just of *what* her image is.

Some of the parody is subtle and hard to tie down for textual analysis, but some, such as the references to Marilyn Monroe and the musicals she often starred in, is more obvious. The subtler parody lies in the knowing way in which Madonna uses the camera, mocking the conventional representations of female sexuality at the same time she conforms to them. Even *Playboy* recognizes her self-parody: "The voice and the body are her bona fides, but Madonna's secret may be her satirical bite. She knows a lot of this image stuff is bullshit: she knows that *you* know. So long as we're all in on the act together, let's enjoy it."[28] One of her former lovers supports this: "Her image is that of a tart, but I believe it's all contrived. She only pretends to be a gold digger. Remember, I have seen the other side of Madonna."[29]

Madonna knows she is putting on a performance. The fact that this knowingness is part of the performance enables the viewer to answer a different interpellation from that proposed by the dominant ideology, and thus to occupy a resisting subject position. The sensitive man watching her material girl performance knows, as she does—as we might do—that this is only a performance. Those who take the performance at face value, who miss its self-parody, are hailed either as ideological subjects in patriarchy, or else they reject the hailing, deny the pleasure, and refuse the communication:

> The *National Enquirer*, a weekly magazine devoted to prurient gossip, quotes two academic psychiatrists denouncing her for advocating teenage promiscuity, promoting a lust for money and materialism, and contributing to the deterioration of the family. Feminists accuse her of revisionism, of resurrecting the manipulative female who survives by coquetry and artifice. "Tell Gloria (Steinem) and the gang," she retorts, "to lighten up, get a sense of humour. And look at my video that goes with Material Girl. The guy who gets me in the end is the sensitive one with no money."[30]

Madonna consistently parodies conventional representations of women, and parody can be an effective device for interrogating the dominant ideology. It takes the defining features of its object, exaggerates and mocks them, and thus mocks those who "fall" for its ideological effect. But Madonna's parody goes further than this; she parodies, not just the stereotypes, but the way in which they are made. She represents herself as one who is in control of her own image and of the process of making it. This, at the reading end of the semiotic process, allows the reader similar control over her own meanings. Madonna's excess of jewelry, of makeup, of trash in her style, offer similar scope to the reader. Excessiveness invites the reader to question ideology: too much lipstick interrogates the tastefully made-up mouth, too much jewelry questions the role of female decorations in patriarchy. Excess overspills ideological control and offers scope for resistance. Thus Madonna's excessively sexual pouting and lipstick can be read to mean that she looks like that not because patriarchy determines that she should, but because she knowingly chooses to. She wears religious icons (and uses a religious name) not to support or attack Christianity's role in patriarchy (and capitalism) but because she chooses to see them as beautiful, sexy ornaments. She makes her own meanings out of the symbolic systems available to her, and in using *their* signifiers and rejecting or mocking *their* signifieds, she is demonstrating *her* ability to make *her own* meanings.

The video of "In the Groove" demonstrates this process clearly. The song is the theme song of *Desperately Seeking Susan,* and the video is a montage of shots from the film. The film is primarily about women's struggle to create and control their own identity in contemporary society, and in so doing to shape the sort of relationships they have with men. Viewers of the video who have seen the film will find plenty of references that can activate these meanings, but the video can also be read as promoting the Madonna look and style. She takes items of urban living, prizes them free from their original social, and therefore signifying, context, and combines them in new ways, and in a new context that denies their original meaning. Thus the crucifix is torn from its religious context and lacy gloves from their context of bourgeois respectability—or, conversely, of the brothel. The dyed blonde hair with the dark roots deliberately displayed is no longer the sign of the tarty slut, and the garter belt and stockings no longer signify soft porn or male kinkiness.

This wrenching of the products of capitalism from their original context and recycling them into a new style is, as Chambers has pointed out, a typical practice of urban popular culture.[31] The products are purified into signifiers; their ideological signifieds are dumped and left behind in their original context. These freed signifiers do not necessarily mean *something,* they do not acquire new signifieds. Rather, the act of freeing them from their ideological context signifies their users' freedom from that context. It signifies the power (however hard the struggle to attain it) of the subordinate to exert some control in the cultural process of making meanings.

The women in *Desperately Seeking Susan* who are struggling to control their social identity and relationships are participating in the same process as subcultures are when they recycle the products of the bourgeoisie to create a style that is theirs—a style that rejects meaning and in this rejection asserts the power of the subordinate to free themselves from the ideology that the meaning bears.

Madonna's videos constantly refer to the production of the image, and they make her control over its production part of the image itself. This emphasis on the making of the image allows, or even invites, an equivalent control by the reader over its reception. It enables girls to see that the meanings of feminine sexuality *can* be in their control, *can* be made in their interests, and that their subjectivities are not necessarily totally determined by the dominant patriarchy.

The constant puns in Madonna lyrics work in a similar way. Puns arise when one word occurs in two discourses—in the case of "Material Girl," those of economics and sexuality: one signifier has simultaneous but different signifieds according to its discourse. The most obvious puns are "give me proper credit," "raise my interest," "experience has made me rich." Less obvious ones are "the boy with the cold hard cash," or "only boys that save their pennies make my rainy day" ("make" has only vestigial sexual meanings, and the homonym between "pennies" and "penis" is only faint). The puns perform typical ideological work by equating economic with sexual success, a common strategy of popular culture in patriarchal capitalism. But puns demand active readers and can never fully control the meanings that are provoked by the yoking of disparate discourses. These puns can expose and thus reject, or at least resist, the economic and sexual subordination of women and the way that each is conventionally used to naturalize the other. The first and last verses of the song are:

> Some boys kiss me some boys hug me
> I think they're OK
> If they don't give me proper credit
> I just walk away.
>
> . . .

> Boys may come and boys may go
> And that's all right you see
> Experience has made me rich
> And now they're after me.[32]

The puns here can be used not to naturalize the dual subordination of woman, but to assert woman's ability to achieve to sexual-economic independence. If a body is all that patriarchy allows a woman to be, then at least she can use it in *her* interests, not in men's.

The pun always resists final ideological closure—the potential meanings provoked by the collision of different discourses is always greater than that proposed by the dominant ideology. Thus "Boy Toy," the name that Madonna has given to her range of products and that the media apply to her, can be read as *Playboy* does when it calls her the "world's number one Boy Toy" or "the compleat Boy Toy."[33] In this reading Madonna is the toy for boys, but the pun can also mean the opposite—that the boy is the toy for her—as she toys with the men in "Material Girl."

Puns are also at work in the word "material," which is located in the discourse of economic capitalism, but which is often used to criticize that discourse either from or a religious view point or from one of a "finer sensibility." In rejecting the materialism of the song, Madonna may be read as proposing the values of a finer sensitivity, a more spiritual love, either secular-erotic or religious-erotic. Madonna's combining of secular and religious love makes explicit a powerful undercurrent of patriarchal Christianity in general—and Catholicism in particular—that traditionally has tried to mobilize man's lustful love for Mary Magdalene, displace it onto Mary the Virgin, and spiritualize it in the process. With Madonna, however, the dualism of the love is denied; it does not fit an either-or dichotomy in which one sort of love is morally superior to the other. By denying the opposition and the moral hierarchy inscribed in it, she rejects the traditional patriarchal Christian evaluation of love and allows sexual or sentimental love to appear on the same level as religious love—certainly not as inferior to it. Her use of the cross as a beautiful ornament for the female body and her construction of nuns as sexy are all part of her critical interrogation of a patriarchal Christian tradition that makes sense of love by means of a moralistic opposition between the spirituality of the virgin and the lust of the whore. Similarly the video of "Like a Virgin" refuses to allow the viewer a moral choice between the white-robed, virginal Madonna bride and the black, sexy Madonna singer. But some readers of "Material Girl" may be less concerned with the relationship between materialism, religious, and sentimental love than with Madonna's ability to choose and control her meanings of religious or secular love, sexuality, and economics.

The video of "Burning Up" is similar in many ways to that of "Material Girl." Both have an "undermining" ending; both employ a consistent series of puns; and both exhibit this knowing parodic excess that is so crucial to the Madonna style. The narrative of "Burning Up" shows Madonna, in a white dress, writhing on a road as she sings of her helpless passion for the uncaring lover who is driving toward her in a car, presumably to run her down. Her love for the boy makes her as helpless a victim as the stereotyped female tied to the railway track in many a silent movie. But the last shot of all shows her in the driver's seat of the car, a knowing, defiant half-smile on her lips, with the boy nowhere to be seen. The narrative denial of female helplessness runs throughout the performance as a countertext to the words of the lyric. So when she sings, "Do you want to see me down on my knees? I'm bending over backward, now would you be pleased," she kneels on the road in front of the advancing car, then turns to throw her head back, exposing her throat in the ultimate posture of submission. But her tone of voice and her look at the camera as she sings have a hardness and a defiance about them that contradict the submissiveness of her body posture and turn the question into a challenge—if not a threat— to the male.

The puns are more subdued and less balanced than those in "Material Girl." The two discourses here are those of sex and religion. Sex may be given a greater emphasis in the text, but the discourse of religion is not far below the surface as she sings of kneeling and burning, of her lack of shame and the something in her heart that just won't die. This yoking of sexuality and religion appears to be performing the traditional ideological work of using the subordination and powerlessness of women in Christianity to naturalize their equally submissive position in patriarchy. But, as in "Material Girl," the text provides the reader with ample opportunities to undermine the dominant ideology while wryly recognizing its presence in the representation, for again the representation of women's sexuality includes the means of that representation and therefore questions its ideological effectivity.

The introductory sequence exhibits this clearly. In the thirty-three seconds before Madonna is shown singing, there are twenty-one shots:

1. Female eye, opening
2. White flowers, one lights up
3. Female mouth, made-up (probably Madonna's)
4. Blue car, lights go on
5. Madonna in white lying on road
6. Male Grecian statue with blank eyes
7. Goldfish in bowl
8. Close-up of male statue, eyes light up
9. Midshot of statue, eyes still lit up
10. Extreme close-up of eye of statue, still lit up
11. Chain round female neck, tightened so that it pinches the flesh
12. Blurred close-up of Madonna with the chain swinging loose
13. Laser beam, which strikes heavy bangles, manacle-like on female wrist
14. Laser beam on goldfish in bowl
15. Madonna removing dark glasses, looking straight at camera
16. Madonna sitting on road
17. Madonna removing dark glasses
18. Madonna lying on her back on the road
19. The dark glasses on the road, an eye appears in one lens, greenish electronic effects merge to realistic image of eye
20. Madonna sitting on road facing camera
21. Close-up of Madonna on road, tilting her head back.

This sequence has two main types of image—ones of looking and ones of subordination or bondage. Traditionally, as the eyes of the Greek statue tell us, looking has been a major way by which men exercise power over women, and the resulting female subordination is shown by Madonna's submissive postures on the road. The goldfish caught in the bowl is an ironic metaphor for the woman held in the male gaze. But the laser beam is a modern "look"; it is impersonal, unlike the traditional male eye beam, and can cut the female free from her bonding manacles, free the goldfish from the bowl. Similarly Madonna's singing frees the chain that has previously been tightened around her neck. Later in the video, as Madonna sings of wanting her lover and wanting to know what she has to do to win him, she tightens and then loosens the same chain about her neck. The next shot is a collage of male eyes into which Madonna's lips are inserted as she sings. Her performance shows how women can be free from the look and the power of the male. Removing the dark glasses as she looks at us is a sign of her control of the look—we see what she allows us to. The glasses replace the earlier image of Madonna lying in the road, but they substitute for her apparent submissiveness an active, electronic, all-seeing eye. Similarily, the video of "Lucky Star" opens and closes with her lowering and raising dark glasses as she looks at the camera, again controlling what we see. In "Borderline," the male photographer is a recurring image as Madonna parodies the photographic model she once was while singing of her desire for freedom. The resulting photograph is shown on the cover of a glossy magazine (called *Gloss!*) being admired by men.

Madonna knows well the importance of the look. This is a complex concept, for it includes how she looks (what she looks like), how she looks (how she gazes at others, the camera in particular), and how others look at her. Traditionally, looking has been in the control of men. Freud even suggests it is an essentially masculine way of exerting control through an extension of voyeurism. But Madonna wrests this control from the male and shows that women's control of the look (in all three senses) is crucial to their gaining control over their meanings within patriarchy.

The ideological effectivity of this is shown in a student essay:

> There is also a sense of pleasure, at least for me and perhaps a large number of other women, in Madonna's defiant look or gaze. In "Lucky Star" at one point in the dance sequence Madonna dances side on to the camera, looking provocative. For an instant we glimpse her tongue: the expectation is that she is about to lick her lips in a sexual invitation. The expectation is denied and Madonna appears to tuck her tongue back into her cheek. This, it seems, is how most of her dancing and grovelling in front of the camera is meant to be taken. She is setting up the sexual idolization of women. For a woman who has experienced this victimization, this setup is most enjoyable and pleasurable, while the male position of voyeur is displaced into uncertainty.[34]

What I have tried to do in this chapter is to demonstrate some of the methodology and theoretical implications of British cultural studies. I shall now try to summarize these.

The television text is a potential of meanings. These meanings are activated by different readers in their different social situations. Because the television text is produced by a capitalist institution, it necessarily bears that ideology. Any subcultural or resistant meanings that are made from it are not "independent" but are made in relation to the dominant ideology. As subcultures are related in various ways to the social system, they will produce an equivalent variety of ways of relating their subcultural meanings of television to those preferred by the dominant ideology. Social relations in capitalism always involve a political dimension (because all such relations are determined more or less directly by the unequal distribution of power), so all meanings arise, in part, from a political base. For some the politics will be those of acceptance, for others, those of rejection or opposition, but for most the politics will be a base for resistance and for the negotiation of meaning.

Cultural analysis can help us to reveal the way the television text can serve as an arena for this struggle for meanings. It treats television as part of the total cultural experience of its viewers. This means that the meanings of television are always intertextual, for it is always read in the context of the other texts that make up this cultural experience. These intertextual relations may be explicit and close, or implicit and tenuous. *Magnum, P.I.* shares many generic characteristics with *The A-Team* but also bears less obvious, though not necessarily less significant, relations with the Vietnam veterans' parade in New York ten years after the war ended and the unveiling of the Vietnam Veterans' Memorial in Washington.

Critical and journalistic comments on television programs, fan magazines, and gossip publications are examples of other types of significant intertextuality. Criticism is, according to Tony Bennett, a series of ideological bids for the meaning of a text, and studying which interpretations are preferred in which publications, for which audiences, can help us to understand why and how certain meanings of the text are activated rather than others.[35] We must be able to understand how that bundle of meanings that we call "Madonna" allows a *Playboy* reader to activate meanings of "the compleat Boy Toy" at the same time a female fan sees her as sexy but not needing men, as being there "all by herself." Publications reflect the meanings that are circulating in the culture, and that will be read back into the television text as an inevitable part of the assimilation of that text into the total cultural experience of the reader.

For culture is a process of making meanings that people actively participate in; it is not a set of preformed meanings handed down to and imposed upon the people. Of course, our "freedom" to make meanings that suit our interests is as circumscribed as any other "freedom" in society—the mass-produced text is produced and circulated by capitalist institutions for economic gain, and is therefore imprinted with capitalist ideology. But the mass-produced text can only be made into a *popular* text by the people, and this transformation occurs when the various subcultures can activate sets of meanings from it and insert these meanings into their daily cultural experience. They take mass-produced signifiers and, by a process of "excorporation" use them to articulate and circulate subcultural meanings.[36]

Gossip is one important means of this active circulation of meanings. The "uses and gratifications" theorists of the 1970s recognized how commonly television was used as a "coin of social exchange," that is, as something to talk about in school yards, suburban coffee mornings, coffee breaks at work, and the family living room.[37] Dorothy Hobson has shown how important gossip is among soap opera fans, and Christine Geraghty has called it the "social cement" that binds the narrative strands of the soap opera together and that binds fans to each other and to the television text.[38] This use of television as a cultural enabler, as a means of participating in the circulation of meanings, is only just becoming clear, and gossip or talk about television is no longer seen as the end in itself (as it was in the "uses and gratifications" approach), but rather as a way of participating actively in that process of the production and circulation of meanings that constitutes culture.

The cultural analysis of television, then, requires us to study three levels of "texts" and the relations between them. First, there is the primary text on the television screen that is produced by the culture industry and that needs to be seen in its context as part of that industry's total production. *Magnum, P.I.* is generically intertextual with *The A-Team* (and with *Cagney and Lacey,* with which it is contrasted), Madonna is an intertextual conglomerate of television, film, record, radio, posters, etcetera. Second, there is a sublevel of texts also produced by the culture industry, though sometimes by different parts of it. These include studio publicity, television criticism and comment, feature articles about shows and their stars, gossip columns, fan magazines, and so on. They can provide evidence of the ways that the various meanings of the primary text are activated and inserted into the culture for various audiences or subcultures. On the third level of textuality lie those texts that the viewers produce themselves: their talk about television; their letters to the papers or magazines; their adoption of styles of dress, of speech, or even of thought into their lives.

These three levels leak into each other. Some secondary texts, such as those of official publicity and public relations, are very close to primary texts; others, such as independent criticism and comment, attempt to "speak for" the third level. Underlying all this, we can, I think, see an oral popular culture adapting its earlier role to one that fits within a mass society. Despite the cultural pessimism of the Frankfurt School, despite the power of ideology to reproduce itself in its subjects, despite the hegemonic force of the dominant classes, the people still manage to make their own meanings and to construct their own culture within, and often against, that which the industry provides for them. Cultural studies aims to understand and encourage this cultural democracy at work.

NOTES

1. Stuart Hall, "The Narrative Construction of Reality," *Southern Review* 17 (1984): 1–17.
2. Angela McRobbie, "Dance and Social Fantasy," in *Gender and Generation*, ed. Angela McRobbie and Mica Nava (London: Macmillan, 1984), pp. 130–61.
3. Annette Kuhn, *Women's Pictures: Feminism and Cinema* (London: Routledge and Kegan Paul, 1982).
4. Louis Althusser, "Ideology and Ideological State Apparatuses," in *Lenin and Philosophy and Other Essays* (London: New Left Books, 1971), pp. 127–86.
5. Stuart Hall et al., "The Unity of Current Affairs Television," in *Popular Television and Film: A Reader*, ed. Tony Bennett et al. (London: British Film Institute/Open University Press, 1981), pp. 88–117; Stephen Heath and Gillian Skirrow, "Television: A World in Action," *Screen* 18, no. 2 (1977): 7–59; John Fiske, "Television and Popular Culture: Reflections on British and Australian Critical Practice," *Critical Studies in Mass Communication* 3, no. 3 (1986): 200–216.
6. Stuart Hall, "Encoding/Decoding," in *Culture, Media, Language*, ed. Stuart Hall et al. (London: Hutchinson, 1980), pp. 128–39.
7. Shere Hite, *The Hite Report on Male Sexuality* (London: Macdonald, 1981).
8. Andrew Moye, "Pornography," in *The Sexuality of Men*, ed. Adrian Metcalf and Martin Humphries (London: Macmillan, 1985).

9. Horace Newcomb, "On the Dialogic Aspect of Mass Communication," *Critical Studies in Mass Communication* 1, no. 1 (1984): 34–50.

10. Ibid.

11. David Morley, *The "Nationwide" Audience: Structure and Decoding* (London: British Film Institute, 1980); Dorothy Hobson, *"Crossroads": The Drama of a Soap Opera* (London: Methuen, 1982); McRobbie, "Dance and Social Fantasy"; Robert Hodge and David Tripp, *Children and Television* (Cambridge: Polity, 1986).

12. Morley, *The "Nationwide" Audience.*

13. Ibid.; see also Charlotte Brunsdon and David Morley, *Everyday Television: "Nationwide"* (London: British Film Institute, 1978).

14. Raymond Williams, *Television: Technology and Cultural Form* (London: Fontana, 1974); John Hartley, "Television and the Power of Dirt," *Australian Journal of Cultural Studies* 1, no. 2 (1983): 68–82.

15. McRobbie, "Dance and Social Fantasy."

16. *Countdown Annual*, 1985, p. 2.

17. Stuart Hall, "Notes on Deconstructing the Popular," in *People's History and Socialist Theory*, ed. Raphael Samuel (London: Routledge and Kegan Paul, 1981).

18. *Countdown*, December 1985, p. 35.

19. See Roland Barthes, *Mythologies* (London: Paladin, 1973); John Fiske, *Introduction to Communication Studies* (London: Methuen, 1982); John Fiske and John Hartley, *Reading Television* (London: Methuen, 1978).

20. Interview by John Fiske, December 1985.

21. *Time*, 27 May 1985, p. 47.

22. *Countdown*, December 1985, p. 70.

23. *Time*, 27 May 1985, p. 47.

24. *Countdown Annual*, 1985, p. 109.

25. Madonna, quoted in *National Times*, 23/29 August 1985, p. 9.

26. Ibid., p. 10.

27. *Playboy*, September 1985, p. 122.

28. Ibid., p. 127.

29. Professor Chris Flynn, quoted in *New Idea*, 11 January 1986, p. 4.

30. *National Times*, 23/29 August 1985, p. 10.

31. Iain Chambers, *Popular Culture: The Metropolitan Experience* (London: Methuen, 1986), pp. 7–13.

32. From "Material Girl," lyrics by Peter Brown and Robert Raus (Minong Publishing Company, B.M.I., 1985).

33. *Playboy*, September 1985, pp. 122, 127.

34. Robyn Blair, student paper, School of Communication and Cultural Studies, Western Australian Institute of Technology, November 1985.

35. Tony Bennett, "The Bond Phenomenon: Theorising a Popular Hero," *Southern Review* 16, no. 2 (1983): 195–225.

36. Lawrence Grossberg, "Another Boring Day in Paradise: Rock and Roll and the Empowerment of Everyday Life," *Popular Music* 4 (1984): 225–57.

37. Denis McQuail et al., "The Television Audience: A Revised Perspective,"

in *The Sociology of Mass Communications*, ed. Denis McQuail (Harmondsworth: Penguin, 1972), pp. 135–65.

38. Hobson, *"Crossroads"*; Christine Geraghty, "The Continuous Serial—a Definition," in *Coronation Street*, ed. Richard Dyer et al. (London: British Film Institute, 1981), pp. 9–26.

FOR FURTHER READING

Two collections of essays provide a good example of work in the British cultural studies tradition; both are strongly recommended: Stuart Hall, Dorothy Hobson, Andrew Lowe, and Paul Willis, eds., *Culture, Media, Language* (London: Hutchinson, 1980)—a collection of essays from the Birmingham Centre for Contemporary Cultural Studies providing good examples of the Marxist, structuralist, and ethnographic approaches to this field; Tony Bennett, Susan Boyd-Bowman, Colin Mercer, and Janet Woollacott, eds., *Popular Television and Film: A Reader* (London: British Film Institute/Open University Press, 1981)—these articles are mainly from *Screen* and provide the best example of this inflection of cultural studies in Britain.

A general overview of work in British cultural studies is given in John Fiske, "Television and Popular Culture: Reflections on British and Australian Critical Practice," *Critical Studies in Mass Communication* 3, no. 3 (1986): 200–216.

The earliest book specifically on television in this tradition is Raymond Williams, *Television: Technology and Cultural Form* (London: Fontana, 1974). Its historical overview argues that cultural needs determine technological development, and its contemporary analysis attempts, sometimes a bit uncertainly, to clarify television's cultural role. John Fiske and John Hartley, *Reading Television* (London: Methuen, 1978) brings European semiotics, particularly the work of Barthes, to bear upon television and links this to British cultural studies. The approach is less historical than Williams's, but the authors give more detailed analysis of programs. Roger Silverstone, *The Message of Television: Myth and Narrative in Contemporary Culture* (London: Heinemann, 1981) gives a very detailed theoretical analysis, drawing largely upon Levi-Strauss, of a thirteen-part TV miniseries; the book is more anthropological and less political than Williams or Fiske and Hartley. John Fiske, "Television Culture: A Critical Theory and Practice," (forthcoming) gives the most up-to-date account of the culturalist approach to television.

Two introductory and/or reference books in this area are John Fiske, *Introduction to Communication Studies* (London: Methuen, 1982); and Tim O'Sullivan, John Hartley, Danny Saunders, and John Fiske, *Key Concepts in Communication* (London: Methuen, 1983). Fiske gives a basic introduction to semiotic theory, definitions of some of its central concepts, and examples of its use. O'Sullivan et al. provide a series of short essays on a wider range of concepts in cultural and structuralist theory. Both these books are in Methuen's Studies in Communication series. Also a part of the series is John Hartley,

Understanding News (London: Methuen, 1983), which shows how the news makes meanings of events, and how these meanings are located within the social system.

Two good examples of ideological analysis of television programs are Stephen Heath and Gillian Skirrow, "Television: A World in Action," *Screen* 18, no. 2 (1977): 7–59; and Stuart Hall, Ian Connell, and Lydia Curti, "The Unity in Current Affairs Television," in Bennett et al., *Popular Television and Film.* The first shows the close analysis typical of screen theory, with its emphasis on the power of television to make meanings for the viewer and position him or her as a reading subject. The second has an equally detailed analysis, but its theory allows more for negotiated readings. Angela McRobbie, "*Jackie*: An Ideology of Adolescent Femininity," in Bernard Waites, Tony Bennett, and Graham Martin, eds., *Popular Culture: Past and Present*, pp. 263–83 (London: Croom Helm, 1982) gives another excellent example of this school of ideological analysis applied not to television but to teenage girls' magazines—well worth reading.

Charlotte Brunsdon and David Morley, *Everyday Television: "Nationwide"* (London: British Film Institute, 1978) applies discourse theory to a detailed and lively analysis of television's way of addressing and interpellating its audience.

The ethnographic work illustrated in Hall et al., *Culture, Media, Language* is developed more fully by David Morley, *The "Nationwide" Audience: Structure and Decoding* (London: British Film Institute, 1980), using an open interview approach; and Dorothy Hobson, *"Crossroads": The Drama of a Soap Opera* (London: Methuen, 1982), using the participant observer method. Angela McRobbie, "Dance and Social Fantasy," in Angela McRobbie and Mica Nava, eds., *Gender and Generation* (London: Macmillan, 1984) shows how girls can integrate television and film into their general cultural experience.

The British Film Institute publishes an excellent series of television monographs. All are well worth reading in addition to the ones already cited (Morley, *The "Nationwide" Audience*, and Brunsdon and Morley, *Everyday Television: "Nationwide"*), but especially recommended are: Colin MacArthur, *Television and History* (1980); Richard Dyer, Christine Geraghty, Marion Jordan, Tony Lovell, Richard Paterson, and John Stewart, *Coronation Street* (1981); Ed Buscomb, ed., *Football and Television* (1975).

The most influential Marxist theory is found in Louis Althusser, "Ideology and Ideological State Apparatuses," in *Lenin and Philosophy and Other Essays* (London: New Left Books, 1971)—a crucial essay. Antonio Gramsci's work is published in the long "prison notebooks," which are for the advanced student only (Gramsci, *Selections from the Prison Notebooks*, ed. and trans. Quentin Hoare and Geoffrey Nowell-Smith [New York: International Publishers, 1971]). A good selection is available in Tony Bennett, Graham Martin, Colin Mercer, and Janet Woollacott, *Culture, Ideology and Social Process* (London: Batsford/ Open University, 1981), which also contains essays commenting on and applying his theory. The book also has an excellent selection of essays on structuralist and cultural theory, though not applied specifically to television.

TELEVISION CRITICISM
A SELECTIVE BIBLIOGRAPHY

ROBERT C. ALLEN, JANE DESMOND, AND GINGER WALSH

ARTICLES

Alvarado, Manuel. "Teaching Television." *Screen Education* 31 (1979): 25–28.

Aufderheide, Pat. "Music Videos: The Look of the Sound." *Journal of Communication* 36 (1986): 57–78.

Barker, David. "Television Production Techniques as Communication." *Critical Studies in Mass Communications* 2 (1985): 234–46.

Baron, Dennis E. "Against Interpretation: The Linguistic Structure of American Drama." *Journal of Popular Culture* 7 (1974): 946–54.

Bazalgette, Cary. "Reagan and Carter, Kojak and Crocker, Batman and Robin?" *Screen Education* 20 (1976): 54–65.

Bazalgette, Cary, and Paterson, Richard. "Real Entertainment: The Iranian Embassy Siege." *Screen Education* 37 (Winter 1980/81)): 55–67.

Ben-Horin, Daniel. "Television Without Tears: An Outline of the Socialist Approach to Popular Television." *Socialist Revolution* 7 (1977): 7–35.

Berger, Arthur Asa. "The Hidden Compulsion in Television." *Journal of the University Film Association* 30 (1978): 41–46.

Black, Peter. "Can One Person Criticise the Full Range of Television?" *Journal of the Society of Film and Television Arts* 2 (1973): 4–5.

Blair, Karin. "The Garden in the Machine: The Why of *Star Trek*." *Journal of Popular Culture* 13 (1979): 310–20.

Boddy, William. "Entering *The Twilight Zone*." *Screen* 25 (1984): 98–108.

Boyd, Douglas A. "The Janus Effect? Imported Television Entertainment Programming in Developing Countries." *Critical Studies in Mass Communications* 1 (1984): 379–91.

Boyd-Bowman, Susan. "*The Day After*: Representations of the Nuclear Holocaust." *Screen* 25, no. 4/5 (July/October 1984): 71–97.

———. "The MTM Phenomenon." *Screen* 26 (1985): 75–87.

Branston, Gill. "TV as Institution." *Screen* 25 (1984): 85–94.

Breen, Myles, and Corcoran, Farrel. "Myth in the Television Discourse." *Communication Monographs* 49 (1982): 127–36.

Browne, Nick. "The Political Economy of the Television (Super)Text." *Quarterly Review of Film Studies* 9 (Summer 1984): 174–82.

Bruck, Peter. "The Social Production of Texts: On the Relation Production/ Product in the News Media." *Communication-Information* 4 (1982): 92–124.

Brunsdon, Charlotte. "*Crossroads*: Notes on Soap Opera." *Screen* 22 (1981): 32–37.

Bryant, John. "Emma, Lucy and the American Situation Comedy of Manners." *Journal of Popular Culture* 13 (1979): 248–55.

Burton, Humphrey. "Criticism at the Receiving End." *Journal of the Society of Film and Television Arts* 2 (1973): 12–14.

Buscombe, Edward. "British Broadcasting and International Communications—An Introduction." *Screen* 24 (1983): 4–5.

———. "Creativity in Television." *Screen Education* 35 (1980): 5–18.

———. "*The Sweeny*—Better Than Nothing?" *Screen Education* 20 (1976): 66–69.

Butler, Jeremy. "Notes on the Soap Opera Apparatus: Televisual Style and *As The World Turns*." *Cinema Journal* 25 (Spring 1986): 53–70.

Carey, John. "A Primer on Interactive Television." *Journal of the University Film Association* 30 (1978): 35–40.

Carpenter, Richard. "*I Spy* and *Mission: Impossible*: Gimmicks and a Fairy Tale." *Journal of Popular Culture* 1 (1967): 286–90.

———. "Ritual, Aesthetics and TV." *Journal of Popular Culture* 3 (1969): 251–59.

Caughie, John. "Progressive Television and Documentary Drama." *Screen* 21 (1980): 9–35.

———. "Rhetoric, Pleasure, and 'Art Television'—*Dreams of Leaving*." *Screen* 22 (1981): 9–31.

———. "Television Criticism." *Screen* 25 (1984): 109–21.

Cawelti, John. "Beatles, Batman and the New Aesthetic." *Midway* 9 (1968): 49–70.

Charland, Maurice. "The Private Eye: From Print to Television." *Journal of Popular Culture* 12 (1979): 210–15.

Cohn, William H. "History for the Masses: Television Portrays the Past." *Journal of Popular Culture* 10 (1976): 280–89.

Collet, Jean. "A Good Use of TV: 6 X 2." *Jump Cut* 27 (1982): 61–63.

Colley, Iain, and Darries, Gill. "*Pennies From Heaven*: Music, Image, Text." *Screen Education* 35 (1980): 63–78.

Connell, Ian. "Commercial Broadcasting and the British Left." *Screen* 24 (1983): 70–80.

———. "Monopoly Capitalism and the Media: Definitions and Struggles." In *Politics, Ideology and the State*, edited by Sally Hibbin, pp. 69–98. London: Lawrence and Wisehart, 1978.

———. "Televising 'Terrorism.'" *Screen* 25 (1984): 76–79.

Corcoran, Farrel. "Television as Ideological Apparatus: The Power and the Pleasure." *Critical Studies in Mass Communications* 1 (1984): 131–45.

Cosgrave, Stuart. "Refusing Consent—The *Oi for England* Project." *Screen* 24 (1983): 92–96.

Dahlgren, Peter. "TV News as a Social Relation." *Media, Culture, and Society* 3 (1981): 291–302.

Day-Lewis, Sean. "The Specialization Issue and Other Problems for the Critic." *Journal of the Society of Film and Television Arts* 2 (1973): 6–8.

Deming, Caren J. "*Hill Street Blues* as Narrative." *Critical Studies in Mass Communications* 2 (1985): 1–22.

Deming, Robert H. "Discourse/Talk/Television." *Screen* 26 (1985): 88–92.

———. "The Television Spectator—Subject." *Journal of Film and Video* 37 (1985): 49–63.

Dennington, John, and Tulloch, John. "Cops, Consensus, and Ideology." *Screen Education* 20 (1976): 37–46.

Derry, Charles. "Television Soap Operas: Incest, Bigamy and Fatal Disease." *Journal of the University Film and Video Association* 35 (1983): 4–16.

Doane, Mary Ann. "Misrecognition and Identity." *Cine-Tracts* 3 (Fall 1980): 25–32.

Downing, John. "Communications and Power." *Socialist Review* 15 (1985): 111–27.

Drummond, Phillip. "Structural and Narrative Constraints and Strategies in *The Sweeny.*" *Screen Education* 20 (1976): 15–35.

Dundes, Alan. "Into the End Zone for a Touchdown: A Psychoanalytic Consideration of American Football." In *Interpreting Folklore*, pp. 199–210. Bloomington: Indiana University Press, 1980.

Dyer, Richard. "Victim: Hermeneutic Project." *Film Form* 1 (1975): 6–18.

Dyer, Richard; Lovell, Terry; and McCrindle, Jean. "Soap Opera and Women." *Edinburgh International Television Festival Programme*, 1977, n.p.

Eaton, Mick. "Television Situation Comedy." *Screen* 19 (25): 61–89.

Eaton, Mick, and Neale, Steve. "On the Air." *Screen* 24 (1983): 62–70.

Eco, Umberto. "Can Television Teach?" *Screen Education* 31 (1979): 15–24.

———. "Towards a Semiotic Enquiry into the TV Message." *Working Papers in Cultural Studies* 3 (1972): n.p.

Elliott, Philip. "Uses and Gratifications: A Critique and a Sociological Alternative." Leicester, U.K.: University of Leicester Centre for Mass Communications Research, n.d.

Elliott, Philip; Murdock, Graham; and Schlesinger, Philip. "'Terrorism' and the State: A Case Study of the Discourses of Television." *Media, Culture, and Society* 5 (1983): 155–77.

Ellis, John. "The Institution of the Cinema." *Edinburgh Magazine* (1977): n.p.

Ellison, Mary. "The Manipulating Eye: Black Images in Non-Documentary TV." *Journal of Popular Culture* 18 (1985): 73–80.

Feuer, Jane. "Melodrama, Serial Form, and Television Today." *Screen* 25 (1984): 4–16.

———. "Narrative Form in Television." In *High Theory, Low Culture*, edited by Colin MacCabe, pp. 101–14. Manchester, Eng.: Manchester University Press, 1986.

Fiske, John. "The Semiotics of Television." *Critical Studies in Mass Communications* 2 (1985): 176–83.

———. "Television: The Flow and the Text." *Madog* 1 (1978): 7–14.

———. "Television and Popular Culture: Reflections on British and Australian Critical Practice." *Critical Studies in Mass Communication* 3, no. 3 (1986): 200–216.

Flitterman, Sandy. "Thighs and Whiskers: The Fascination of *Magnum, P.I.*" *Screen* 26, no. 2 (March/April 1985): 42–58.

Forbes, Jill. "Everyone Needs Standards—French Television." *Screen* 24 (1983): 28–39.

Forbes, Jill, and Nice, Richard. "Pandora's Box: Television and the 1978 French General Elections." *Media, Culture, and Society* 1 (1979): 35–50.

Furst, Terry. "Social Change and the Commercialization of Professional Sports." *International Review of Sports Sociology* 6 (1971): 153–70.

Gardner, Carl, and Henry, Margaret. "Racism, Anti-racism and Access Television: The Making of *Open Door*." *Screen Education* 31 (1979): 69–81.

Gardner, Carl, and Sheppard, Julie. "Transforming Television—Part One, the Limits of Left Policy." *Screen* 25 (1984): 26–40.

Garnham, Nicholas. "Public Service Versus the Market." *Screen* 24 (1983): 6–27.

———. "Television Documentary and Ideology." *Screen* 13 (1972): 109–15.

Gibson, William. "Network News: Elements of a Theory." *Social Text* 3 (1980): 88–111.

Gilbert, W. Stephen. "The TV Play: Outside the Consensus." *Screen Education* 35 (1980): 35–44.

Gitlin, Todd. "Media Sociology: The Dominant Paradigm." *Theory and Society* 6 (1978): 205–53.

———. "Spotlights and Shadows: Television and the Culture of Politics." *College English* 38 (1977): 789–801.

Gray, Herman. "Television and the New Black Man: Black Male Images in Prime-Time Situation Comedy." *Media, Culture, and Society* 8 (1986): 223–42.

Grealy, Jim. "Notes on Popular Culture." *Screen Education* 22 (1977): 5–11.

Greenberg, Harvey R. "In Search of Spock: A Psychoanalytic Inquiry." *Journal of Popular Film and Television* 12 (1984): 52–65.

Gutch, Robin. "Whose Telly Anyway?" *Screen* 25 (1984): 122–27.

Hall, Stuart. "Deviancy, Politics and the Media." In *Deviance and Social Control*, edited by P. Rock and M. McIntosh, pp. 261–305. London: Tavistock, 1974.

Halloran, James D. "Understanding Television." *Screen Education* 14 (1975): 4–13.

Hartley, John. "Television and the Power of Dirt." *Australian Journal of Cultural Studies* 1, no. 2 (1983): 68–82.

Hartley, John, and Fiske, John. "Myth—Representation: A Cultural Reading of News at Ten." *Communications Studies Bulletin* 4 (1977): 12–33.

Heath, Stephen, and Skirrow, Gillian. "Television: A World in Action." *Screen* 18, no. 2 (Summer 1977): 7–59.

Herridge, Peter. "Television, the 'Riots', and Research." *Screen* 24 (1983): 86–91.

Hilmes, Michele. "The Television Apparatus: Direct Address." *Journal of Film and Video* 37, no. 4 (Fall 1985): 27–36.

Hoffer, Tom W., and Nelson, Richard Alan. "Docudrama on American Television." *Journal of the University Film Association* 30 (1978): 21–28.

Homans, Peter. "Psychology and Popular Culture: Ideological Reflections on *M*A*S*H*." *Journal of Popular Culture* 17 (1983): 3–21.

Honeyford, Susan. "Women and Television." *Screen* 21, no. 2 (1980): 49–52.

Houston, Beverle. "Viewing Television: The Metapsychology of Endless Consumption." *Quarterly Review of Film Studies* 9, no. 3 (Summer 1984): 183–95.

Hurd, Geoff. "*The Sweeny*—Contradiction and Coherence." *Screen Education* 20 (1976): 47–53.

Kagan, Norman. "Amos 'n' Andy: Twenty Years Late, or Two Decades Early?" *Journal of Popular Culture* 9 (1975): 71–76.

Kaplan, E. Ann. "History, Spectatorship, and Gender Address in Music Television." *Journal of Communication Inquiry* 10, no. 1 (Winter 1986): 3–14.

———. "A Post-Modern Play of the Signifier? Advertising, Pastiche and Schizophrenia in Music Television." In *Television in Transition*, edited by Phillip Drummond and Richard Paterson, pp. 146–63. London: British Film Institute, 1985.

———. "Sexual Difference, Pleasure and the Construction of the Spectator in Music Television." *Oxford Literary Review* 8, no. 1/2 (1986): 113–23.

Kaplan, Frederick I. "Intimacy and Conformity in American Soap Opera." *Journal of Popular Culture* 9 (1975): 622–25.

Kellner, Douglas. "TV, Ideology and Emancipatory Popular Culture." *Socialist Review* 9 (1979): 13–53.

Kerr, Paul. "Situation Comedies . . ." *Screen* 24 (1983): 71–74.

Kervin, Denise. "Reality According to Television News: Pictures from El Salvador." *Wide Angle* 7 (1985): 61–71.

Kinder, Marsha. "Music Video and the Spectator: Television, Ideology, and Dream." *Film Quarterly* 38, no. 1 (Fall 1984): 3–15.

Kreizenbeck, Alan. "Soaps: Promiscuity, Adultery and 'New Improved Cheer.'" *Journal of Popular Culture* 17 (1983): 175–81.

Laing, Dave. "Music Video—Industrial Product, Cultural Form." *Screen* 26 (1985): 78–83.

Langer, John. "Television's 'Personality System.'" *Media, Culture, and Society* 3 (1981): 351–66.

Levinson, Richard M. "From Olive Oyl to Sweet Polly Purebred: Sex Role Stereotypes and Televised Cartoons." *Journal of Popular Culture* 9 (1975): 561–72.

Linick, Anthony. "Britannia Rules the Air Waves: Television Programming in Transatlantic Perspective." *Journal of Popular Culture* 7 (1974): 918–27.

———. "Magic and Identity in Television Programming." *Journal of Popular Culture* 3 (1970): 644–55.

Lopate, Carol. "Daytime Television: You'll Never Want to Leave Home." *Radical America* 2 (1977): 33–51.

Lusted, David. "Feeding the Panic and Breaking the Cycle—'Popular TV and Schoolchildren.'" *Screen* 24 (1983): 81–93.

McAdow, Ron. "Experience of Soap Opera." *Journal of Popular Culture* 7 (1974): 955–65.

McArthur, Colin. "Point of Review." *Screen Education* 35 (1980): 59–62.

MacDonald, J. Fred. "Black Perimeters—Paul Robeson, Nat King Cole and the Role of Blacks in American TV." *Journal of Popular Film and Television* 7 (1979): 246–64.

———. "The Cold War as Entertainment in 'Fifties Television." *Journal of Popular Film and Television* 7 (1978): 3–31.

McGrath, John. "TV Drama: The Case Against Naturalism." *Sight and Sound* 46 (1977): 100–105.

McKinley, Robert. "Culture Meets Nature on the Six O'Clock News: American Cosmology." *Journal of Popular Culture* 17 (1983): 109–14.

Mander, Mary. "*Dallas*: The Mythology of Crime and Moral Occult." *Journal of Popular Culture* 17 (1983): 44–50.

Manvell, Roger. "Why Television Criticism Differs From Other Forms of Criticism." *Journal of the Society of Film and Television Arts* 2 (1973): 1–3.

Mellencamp, Patricia. "Situation and Simulation: An Introduction to *I Love Lucy*." *Screen* 26 (1985): 30–40.

Merlman, Richard. "Power and Community in Television." *Journal of Popular Culture* 2 (1968): 63–80.

Mills, Adam, and Rice, Phil. "Quizzing the Popular." *Screen Education* 41, no. 4 (Winter/Spring 1982): 15–25.

Modleski, Tania. "The Search for Tomorrow in Today's Soap Operas: Notes on a Feminine Narrative Form." *Film Quarterly* 33 (1979): 12–21.

Mohr, Howard. "TV Weather Programs." *Journal of Popular Culture* 4 (1971): 628–33.

Montgomery, Kathryn. "Television Criticism: An Integrative Approach." In *College Course Files*, edited by Patricia Erens, pp. 49–54. University Film and Video Association, 1985.

———. "Writing About Television in the Popular Press." *Critical Studies in Mass Communications* 2 (1985): 74–89.

Morse, Margaret. "Talk, Talk, Talk—the Space of Discourse in Television." *Screen* 26, no. 2 (March/April 1985): 2–15.

Mould, David H. "Historical Trends in the Criticism of the Newsreel and Television News, 1930–1955." *Journal of Popular Film and Television* 12 (1984): 118–26.

Murdock, Graham. "Authorship and Organisation." *Screen Education* 35 (1980): 19–34.

Newcomb, Horace M. "American Television Criticism, 1970–1985." *Critical Studies in Mass Communications* 3 (1986): 217–28.

———. "On the Dialogic Aspects of Mass Communication." *Critical Studies in Mass Communication* 1, no. 1 (1984): 34–50.

Newcomb, Horace, and Hirsch, Paul M. "Television as a Cultural Forum: Implications for Research." *Quarterly Review of Film Studies* 8, no. 3 (1983): 45–56.

Norton, Suzanne Frentz. "Tea Time on the 'Telly': British and Australian Soap Opera." *Journal of Popular Culture* 19 (1985): 3–19.

Nowell-Smith, Geoffrey. "Television—Football—The World." *Screen* 19 (25): 45–59.

Oakley, Giles. "Cinematic Comparisons." *Screen* 24 (1983): 81–85.

Page, Malcolm. "The British Television Play: A Review Article." *Journal of Popular Culture* 5 (1972): 806–20.

Paterson, Richard. "Planning the Family: The Art of the Television Schedule." *Screen Education* 35 (1980): 79–86.

———. "*The Sweeny*." *Screen Education* 20 (1976): 5–14.

Pearson, Tony. "Teaching Television." *Screen* 24 (1983): 35–43.

Petro, Patrice. "Mass Culture and the Feminine: The 'Place' of Television in Film Studies." *Cinema Journal* 25 (1986): 5–21.

Poole, Michael. "The Cult of the Generalist: British Television Criticism 1936–83." *Screen* 25, no. 2 (March/April 1984): 41–62.

Porter, Dennis. "Soap Time: Thoughts on a Commodity Art Form." *College English* 38 (1977): 782–88.

Porter, Vincent. "Video Recording and the Teacher." *Screen Education* 35 (1980): 87–90.

Pringle, Ashley. "A Methodology for Television Analysis with Reference to the Drama Series." *Screen* 13 (1972): 117–28.

Requena, Jesus G. "Narrativity/Discursivity in the American Television Film." *Screen* 22 (1981): 38–42.

Robinson, Lillian. "What's My Line? Telefiction and Women's Work." In *Sex, Class and Culture*, pp. 310–44. Bloomington: Indiana University Press, 1978.

Sahin, Haluk. "Ideology of Television: Theoretical Framework and a Case Study." *Media, Culture, and Society* 1 (1979): 161–70.

Sahin, Haluk, and Robinson, J. P. "Beyond the Realm of Necessity: Television and the Colonization of Leisure." *Media, Culture, and Society* 3 (1981): 85–96.

Scannell, Paddy. "The Social Eye of Television, 1946–1955." *Media, Culture, and Society* 1 (1979): 97–106.

Schulze, Laurie Jane. "*Getting Physical*: Text/Context/Reading and the Made-For-Television Movie." *Cinema Journal* 25, no. 2 (Winter 1986): 35–50.

Schwichtenberg, Cathy. "*Charlie's Angels* (ABC-TV)." *Jump Cut* 24/25 (1981): 13–15.

———. "*The Love Boat*: The Packaging and Selling of Love, Heterosexual Romance, and Family." *Media, Culture, and Society* 6, no. 3 (July 1984): 301–11.

Seiter, Ellen. "Men, Sex, and Money in Recent Family Melodrama." *Journal of the University Film and Video Association* 35 (1983): 17–27.

———. "Promise and Contradiction: The Daytime Television Serials." *Screen* 23 (1982): 150–63.

———. "The Role of the Woman Reader: Eco's Narrative Theory and Soap Opera." *Tabloid* 6 (1981): 36–43.

Shatzkin, Roger. "*Shogun* (NBC-TV)." *Jump Cut* 24/25 (1981): 16.

Silverstone, Roger. "An Approach to the Structural Analysis of the Television Message." *Screen* 17 (1976): 9–40.

———. "Narrative Strategies in Television Science—A Case Study." *Media, Culture, and Society* 6 (1984): 377–410.

———. "The Right to Speak: On a Poetic for Television Documentary." *Media, Culture, and Society* 5 (1983): 137–54.

Simpson, Philip. "Talking Heads." *Screen* 25 (1984): 80–84.

Smith, Keith. "Viewings: Which, To Whom and For What?" *Journal of the Society of Film and Television Arts* 2 (1973): 10–12.

Spence, Jo. "An *Omnibus* Dossier." *Screen* 24 (1983): 40–52.

Spence, Louise. "Life's Little Problems . . . and Pleasures: An Investigation Into the Narrative Structures of *The Young and the Restless*." *Quarterly Review of Film Studies* 9 (1984): 301–8.

Stein, Howard F. "In Search of *Roots*: An Epic of Origins and Destiny." *Journal of Popular Culture* 11 (1977): 11–17.

Stone, Douglas. "TV Movies and How They Get That Way." *Journal of Popular Film and Television* 7 (1979): 147–49.

Surlin, Stuart. "Television Criticism in Canada." *Critical Studies in Mass Communications* 2 (1985): 80–83.

Thomas, Sari. "Reality, Fiction and Television." *Journal of the University Film Association* 30 (1978): 29–34.

Thompson, John O. "Tragic Flow." *Screen Education* 35 (1980): 45–58.

Thomson, David. "TV Weather." *Sight and Sound* 49 (1980): 87–90.

Tolson, Andrew. "Anecdotal Television." *Screen* 26 (1985): 18–27.

Tomasulo, Frank P. "The Spectator-in-the-Tube: The Rhetoric of Donahue." *Journal of the University Film and Video Association* 36 (1984): 5–12.

Trevino, Jesus Salvador. "Latino Portrayals in Film and Television." *Jump Cut* 30 (1985): 14–16.

Tuchman, Gaye. "Television News and the Metaphor of Myth." *Studies in the Anthropology of Visual Culture* 5 (1978): 56–62.

Tulloch, John. "Gradgrind's Heirs: The Quiz and the Presentation of Knowledge by British Television." *Screen Education* 19 (1976): 3–13.

Tyrrell, William Blake. "*Star Trek* as Myth and Television as Mythmaker." *Journal of Popular Culture* 10 (1977): 711–19.

Verschuure, Eric P. "Stumble, Bumble, Mumble: TV's Image of the South." *Journal of Popular Culture* 16 (1982): 92–96.

Vianello, Robert. "The Power Politics of 'Live' Television." *Journal of Film and Video* 37 (1985): 26–40.

Watson, Mary Ann. "Television Criticism in the Popular Press." *Critical Studies in Mass Communications* 2 (1985): 66–74.

White, Duffield. "Television Non-Fiction as Historical Narrative." *Journal of Popular Culture* 7 (1974): 928–33.

White, Mimi. "Crossing Wavelengths: The Diegetic and Referential Imaginary of American Commercial Television." *Cinema Journal* 25, no. 2 (Winter 1986): 51–64.

———. "Television Genres: Intertextuality." *Journal of Film and Video* 37, no. 3 (1985): 41–47.

Williams, Brien R., and Fulton, Cheryl. "A Study of Visual Style and Creativity in Television." *Journal of the University Film and Video Association* 36 (1984): 23–35.

Williams, Carol Traynor. "It's Not So Much 'You've Come a Long Way Baby'—as 'You're Gonna Make It After All.'" *Journal of Popular Culture* 7 (1974): 981–89.

Williams, Martin. "TV: Tell Me a Story." *Journal of Popular Culture* 7 (1974): 895–99.

Williams, Raymond. "Television and Teaching." *Screen Education* 31 (1979): 5–14.

Woal, Michael B. "Defamiliarization in Television Viewing: Aesthetic and Rhetorical Modes of Experiencing Television." *Journal of the University Film and Video Association* 34 (1982): 25–32.

Wren-Lewis, Justin. "The Encoding/Decoding Model: Criticisms and Redevelopments for Research on Decoding." *Media, Culture, and Society* 5 (1983): 179–97.

———. "TV Coverage of the Riots." *Screen Education* 40 (1981/82): 15–33.

Wright, John L. "TUNE-IN: The Focus of Television Criticism." *Journal of Popular Culture* 7 (1974): 887–94.

Wyver, John. "Screening Television." *Screen* 24 (1983): 75–80.

Zettl, Herbert. "The Rare Case of Television Aesthetics." *Journal of the University Film Association* 30 (1978): 3–8.

Zimmerman, Patricia R. "Good Girls, Bad Women: The Role of Older Women on *Dynasty*." *Journal of Film and Video* 37 (1985): 89–92.

BOOKS

Allen, Robert C. *Speaking of Soap Operas*. Chapel Hill: University of North Carolina Press, 1985.

Ang, Ien. *Watching "Dallas": Soap Opera and the Melodramatic Imagination*. Translated by Della Couling. London: Methuen, 1985.

Arlen, Michael J. *The Camera Age: Essays on Television*. New York: Farrar, Straus and Giroux, 1981.

———. *The Living Room War*. Middlesex, Eng.: Harmondsworth, 1982.

———. *The View From Highway 1: Essays on Television*. New York: Farrar, Straus and Giroux, 1976.

Berger, Arthur Asa. *Television as an Instrument of Terror: Essays on Media, Popular Culture and Everyday Life*. New Brunswick, N.J.: Transaction Books, 1980.

———. *The TV-Guided American*. New York: Walker Publishing Company, 1976.

Brunsdon, Charlotte, and Morley, David. *Everyday Television: "Nationwide"*. London: British Film Institute, 1978.

Bussell, Jan. *The Art of Television*. London: Faber and Faber, 1952.

Cantor, Muriel G., and Pingree, Suzanne. *The Soap Opera*. Beverly Hills, Calif.: Sage, 1983.

Cassata, Mary, and Skill, Thomas. *Life on Daytime Television: Tuning in American Serial Drama*. Norwood, N.J.: Ablex, 1983.

Conrad, Peter. *Television: The Medium and Its Manners*. Boston: Routledge and Kegan Paul, 1982.

Drummond, Phillip, and Paterson, Richard. *Television in Transition*. London: British Film Institute, 1985.

Dyer, Richard. *Light Entertainment*. London: British Film Institute, 1973.

Dyer, Richard; Geraghty, Christine; Jordan, Marion; Lovell, Terry; Paterson, Richard; and Stewart, John. *Coronation Street*. London: British Film Institute, 1981.

Edmonds, Robert. *The Sights and Sounds of Television: How the Aesthetic Experience Influences Our Feelings*. New York: Teachers College Press, 1982.

Elliott, Philip. *The Making of a Television Series*. London: Constable, 1972.

Ellis, John. *Visible Fictions: Cinema, Television, Video*. London: Routledge and Kegan Paul, 1982.

Ellison, Harlan. *The Glass Teat: Essays of Opinion on the Subject of Television*. New York: Ace Books, 1970.

———. *The Other Glass Teat*. New York: Pyramid Books, 1975.

Epstein, Edward J. *News From Nowhere*. New York: Random House, 1974.

Esslin, Martin. *The Age of Television*. San Francisco: W. H. Freeman and Company, 1982.

Feuer, Jane; Kerr, Paul; and Vahimagi, Tise. *MTM: Quality Television*. London: British Film Institute, 1984.

Fiske, John. *Introduction to Communication Studies*. London: Methuen, 1982.

Fiske, John, and Hartley, John. *Reading Television*. London: Methuen, 1978.

Foster, Harold M. *The New Literacy: The Language of Film and Television*. Urbana, Ill.: National Council of Teachers of English, 1979.

Fowles, Jib. *Television Viewers Vs. Media Snobs: What TV Does for People*. New York: Stein and Day, 1982.

Freeman, Don. *Eyes as Big as Cantaloupes: An Irreverent Look at TV*. San Diego, Calif.: Joyce Press, 1978.

———. *In a Flea's Navel: A Critic's Love Affair with Television*. New York: A. S. Barnes and Company, 1980.

Gans, Herbert J. *Popular Culture and High Culture*. New York: Basic Books, 1974.

Garnham, Nicholas. *Structures of Television*. London: British Film Institute, 1978.

Goethals, Gregor T. *The TV Ritual: Worship at the Video Altar*. Boston: Beacon Press, 1981.

Gitlin, Todd. *Inside Prime Time*. New York: Pantheon, 1985.

Glasgow University Media Group. *Bad News*. London: Routledge and Kegan Paul, 1976.

———. *More Bad News*. London: Routledge and Kegan Paul, 1980.

Grote, David. *The End of Comedy: The Sit-Com and the Comedic Tradition*. Hamden, Conn.: Shoe String Press, 1983.

Hall, Stuart, and Whannel, Paddy. *The Popular Arts*. London: Pantheon Books, 1964.

Hall, Stuart; Hobson, Dorothy; Lowe, Andrew; and Willis, Paul. *Culture, Media, Language*. London: Hutchinson, 1980.

Hartley, John. *Understanding News*. London: Methuen, 1983.

Higgins, Anthony Paul. *Talking About Television*. London: British Film Institute, 1966.

Himmelstein, Hal. *On the Small Screen: New Approaches in Television and Video Criticism*. New York: Praeger, 1981.

———. *Television Myth and the American Mind*. New York: Praeger, 1984.

Hobson, Dorothy. *"Crossroads": The Drama of a Soap Opera*. London: Methuen, 1982.

Intintoli, M. *Taking Soaps Seriously*. New York: Praeger, 1985.

Kaminsky, Stuart M., and Mahan, Jeffrey H. *American Television Genres*. Chicago: Nelson-Hall, 1985.

MacDonald, J. Fred. *Blacks and White TV: Afro-Americans in Television and Video Criticism*. Chicago: Nelson-Hall, 1983.

Marc, David. *Demographic Vistas: Television in American Culture*. Philadelphia: University of Pennsylvania Press, 1984.

Mayer, Martin. *About Television*. New York: Harper and Row, 1972.

Meehan, Diane M. *Ladies of the Evening: Women Characters of Prime-Time Television*. Metuchen, N.J.: Scarecrow Press, 1983.

Modleski, Tania. *Loving with a Vengeance: Mass-Produced Fantasies for Women*. London: Methuen, 1982.

Morley, David. *Family Television: Cultural Power and Domestic Leisure*. London: Comedia, 1986.

———. *The "Nationwide" Audience: Structure and Decoding*. London: British Film Institute, 1980.

Newcomb, Horace. *TV: The Most Popular Art*. New York: Anchor, 1974.

Newcomb, Horace, and Adler, Dick. *The Producer's Medium: Conversations with Creators of American TV*. New York: Oxford University Press, 1983.

Poole, Michael, and Wyver, John. *Powerplays: Trevor Griffiths in Television.* London: British Film Institute, 1984.

Silverstone, Roger. *The Message of Television: Myth and Narrative in Contemporary Culture.* London: Heinemann, 1981.

Sklar, Robert. *Prime-Time America.* New York: Oxford University Press, 1980.

Sopkin, Charles. *Seven Glorious Days, Seven Fun-Filled Nights.* New York: Simon and Schuster, 1968.

Thompson, John O. *Monty Python: Complete and Utter Theory of the Grotesque.* London: British Film Institute, 1982.

Tulloch, John, and Alvarado, Manuel. *"Doctor Who": The Unfolding Text.* London: Macmillan, 1983.

Watson, James. *Television in Transition: A Report on the New Electronics Media.* Chicago: Crain Books, 1983.

Wicking, Christopher, and Vahimagi, Tise. *The American Vein: Directors and Direction in Television.* New York: E. P. Dutton, 1979.

Williams, Raymond. *Television: Technology and Cultural Form.* New York: Schocken Books, 1975.

Williamson, Judith. *Decoding Advertisements: Ideology and Meaning in Advertising.* London: Marion Boyars, 1978.

Zettl, Herbert. *Sight, Sound, Motion: Applied Media Aesthetics.* Belmont, Calif.: Wadsworth, 1973.

ANTHOLOGIES

Adler, Richard, and Cater, Douglas, eds. *Television as a Cultural Force.* New York: Praeger, 1975.

Bennett, Tony; Boyd-Bowman, Susan; Mercer, Colin; and Woollacott, Janet, eds. *Popular Television and Film: A Reader.* London: British Film Institute/ Open University Press, 1981.

Brandt, George W., ed. *British Television Drama.* Cambridge: Cambridge University Press, 1981.

Brown, Les, and Walker, Savannah Waring, eds. *Fast Forward: The New Television and American Society Essays From Channels of Communications.* Kansas City: Andrews and McMeel, 1983.

Caughie, John, ed. *Television, Ideology and Exchange.* London: British Film Institute, 1978.

D'Agostino, Peter, ed. *Transmission: Theory and Practice for a New Television Aesthetics.* New York: Tanam Press, 1985.

Davis, Douglas, and Simmons, Allison, eds. *The New Television: A Public/ Private Art.* Cambridge, Mass.: The MIT Press, 1977.

Hazard, Patrick D., ed. *TV as an Art: Some Essays on Criticism.* Champaign, Ill.: National Council of Teachers of English, 1966.

Lowe, Carl, ed. *Television and American Culture.* New York: H. W. Wilson Company, 1981.

Masterman, Len, ed. *Television Mythologies: Stars, Shows and Signs*. London: Comedia/UK Media Press, 1984.

Newcomb, Horace, ed. *Television: The Critical View*. 3d ed. New York: Oxford University Press, 1982.

O'Connor, J. E., ed. *American History/American Television: Interpreting the Video Past*. New York: Frederick Ungar, 1983.

Rowland, Willard D., Jr., and Watkins, Bruce, eds. *Interpreting Television: Current Research Perspectives*. Beverly Hills, Calif.: Sage, 1984.

INDEX